Beyond
PRESCRIPTIONS

Beyond PRESCRIPTIONS

Dr. Rakesh Periwal

PARTRIDGE
A Penguin Random House Company

ISBN:	Hardcover	978-1-4828-4789-5
	Softcover	978-1-4828-4788-8
	eBook	978-1-4828-4790-1

To order additional copies of this book, contact
Partridge India
000 800 10062 62
orders.india@partridgepublishing.com

www.partridgepublishing.com/india

As a doctor, especially as an Indian physician, our work entails a lot of writing. Thus writing a book appeared more like a day to day work, especially as it involved writing all about my experiences as a doctor; except that this work goes beyond writing medical prescriptions.

The human aspects of medicine have been as fascinating as the medical science itself; add to that the pace at which things have moved in the last two decades, stories were sure to be woven! The doctor patient relationship is not just about disease and treatment. There are also two human beings involved and the stories are written to reflect this angle. The domain of my practise, from general medicine to critical care gives me an opportunity to experience this on a wider canvas.

The scientific content warranted some technical details and I have made a conscious effort to keep them short and meaningful. I hope they also provide better insights into the practise of medicine. The stories area a mixed bag of emotions and I believe the reader feels the same intensities as I did living through them.

Reading the book is like spending a little time with me. Thanks for being my virtual company.

CONTENTS

RELATIONSHIP DISEASED

Disease is the most powerful enemy of mankind, it has taken more lives than even warfare; and just like wars, it has repercussions which goes way beyond what is visible. Warfare has redefined civilisation, I have seen disease redefine a human relationship. I was caught in this tussle as an outsider, not even as a doctor but as a mere spectator who had some technical knowledge, and what I saw brought up some of the most intense human emotions. It is about a lethal disease whose first victim is actually a relationship. Not one but many. A tragedy created by the Almighty where the core issue is shifted from survival to surviving a relation. And I say the Almighty here because of the way the script was spread out, not even the most imaginative human mind could have laid it down the way it happened. This is about a beautiful girl who is hit by a disease like a thunderbolt which forever changes her life, not just her body but everything around her.

I had seen this girl in the best of her times. She was beautiful beyond description. I generally pick up everything in a glance; the skin, the hair, the eyes, the lips, the body, the nails, the dress, the shoes, almost all the overt details. But I really found it difficult with her because I got stuck almost everywhere! She is one of those faces you come across sometimes, equally befitted with everything else. My curiosity about her was short lived because I came to know from my cousin that she was her sister in law. She was still single but my interests in her were merely what a connoisseur would have in a piece of art. I was already committed (even that equation could have been reset but I had already ran out of energy in getting myself committed!) and so there was no story to be made. What I didn't realise was that I would be a witness to another story which probably was more powerful than the romance between a man and a woman.

The girl was in her late twenties and as should have happened, she was engaged officially. "The boy loves her a lot", my cousin said. Anyone would, I thought. 'She is very lucky." I was not sure who was more. The marriage was destined for the summers and her in laws insisted that I should be around. An Indian wedding comes with a long countdown and the notification period is long enough to make plans. It would be a great Indian wedding; with an angel looking girl, it would be more like a fairy tale to me!

It was an early morning in Delhi and I was getting ready for work when I received a phone call from my brother in law. "You have to urgently come to The Gangaram Hospital, my sister is admitted and the doctors say she is in need of an urgent surgery. We need you to be around." Off went the call, perhaps he was not interested in the reply. It almost meant be whatever, be there. I immediately called my colleague that I would need an off and I also hung up, because what I intended to say was that I was not coming and please manage. Neither was I interested in a reply. I took a cab and rushed to the hospital. On my way I thought of calling my cousin, at least I should have some idea what was going on.

"Where are you?"

"I have just reached hospital and was about to call you. I hope you are reaching soon. We are eagerly waiting for you." I could understand that they needed someone to talk to the Doctor, understand the urgency of the situation and help them in deciding the next course of things.

"I am on my way. But who actually is sick?" my cousin had three sister in laws.

"My eldest sister in law, the one who was to be married", she clarified in case I did not know the hierarchy.

"But you told me that she had been to her native town for the final wedding preparations. How come she is here?"

"Actually she had a severe headache and then lost consciousness. Thereafter she was immediately rushed to the hospital and the doctor here says she is in need of an urgent surgery."

"Is she conscious now?" I did not want them to waste any time if that was not the case.

"Yes."

"I will be there in an hour."

My neurons were now firing incoherently. The disease; the context in terms of the timing and the person; and me out of nowhere; the one hour drive was getting me restless. I did not know how important my presence would be but I needed to be there, may be for my own reasons.

I reached the hospital and very soon I was with the treating doctor. The introduction was short and perhaps he knew that I might be involved in the decision making. I had picked up the name plate, he was a neurosurgeon. And the sequence of headache, unconsciousness, hospitalisation and emergency surgery; something was seriously wrong, I realised. He immediately took out her MRI plate, placed it on the view box and started.

"She has a Medulloblastoma. She needs an immediate debulking surgery; we can decide the future course of action later."

He was brief and expected me to understand all.

Then I did not know how to read an MRI but the gross distortion of the posterior fossa by the tumour was easily appreciable. Although I could not make much in terms of the actual implications of the MRI findings, one thing that I knew for sure was that the posterior fossa has very little space and even a small tumour can create havoc. It had already created one. The two cerebral hemispheres compresses to a stem like shape which is an area very close to the fossa and also has the important centres of breathing, blood pressure and heart beat regulation and derangements here can be lethal. She was lucky to be surviving.

"What are her chances of survival?"

"I cannot say exactly. We see more of this tumour in the child hood, for which we have a lot of data, not only on survival but also on growth and intellectual development, but in this case she is in her late adulthood and thus there will be little information available."

I knew I was in no position to give any scientific advice, I was rather interested in collecting all the information, summarise and quickly discuss it with the family. "I will get back shortly." I left.

Right outside the room my brother in law was waiting for me. As soon as I came out, he took me to a corner. "Her fiancé is also around. He does not know about the condition but I am told that she has some brain tumour. He wants to talk to you. And I don't know to what extent he should be told."

I could understand that he was only the fiancé and not the husband. What if the marriage was two weeks away, they were still unmarried and that may not necessarily change. Life had come to such a situation. And before I could get back to myself, her fiancé was there in front of me. Only our names were exchanged for introduction.

"What do you say?" he started off immediately.

To me he appeared more curious, more concerned rather than troubled. There seemed to be a genuine sense of enquiry. In fact the person who was in trouble was me. What could I say? I knew very little about the disease, equally little about the girl and the boy. All I knew was how did they all look; the girl, the boy and the tumour. (I had seen the picture of the tumour on the MRI!).

"I mean does she need an urgent surgery?" he was back at me again.

I took a deep breath and started slowly. "She has a growth in the posterior part of her brain which is compressing her normal brain. Surgery could be of survival benefit."

I had worked out the statement in my mind very well and I knew this was all I needed to tell. Growth is a better word to address to a tumour in such circumstances. I was also more particular about the normal part of her brain than the abnormal lesion and truly enough that was of greater concern. I did not say about the risks of the surgery nor gave any approval and pointed it to him that there was an issue of survival (and not of mortality!)

"Then we should not delay it."

I was wondering who the doctor would approach for a consent for surgery. I perhaps was getting carried away. It had to be her father. I told her father that they should go ahead with the surgery as planned. And I was thinking whether that would bring an end to her miseries or would lead to something else even worse. There was so much more to happen, beyond her disease and its treatment.

The operation was successful, but that was only part of the treatment. She would have to receive chemotherapy and radiotherapy for any possible chance of cure. My evening was spent reading the literature on the subject. For some time I had to turn away from the human angle (which in fact was bothering me more at that point) and look at the scientific data. The 5yr survival in children with appropriate and timely treatment was as good as 80%, which was excellent but for those with advanced disease where the tumour had spread to other areas of the nervous system, the outcome was grim. But there was very little to read about the disease in adults. Then how come the doctor reached the conclusion that she has Medulloblastoma? Even without the histological report, which would be available only after the surgically resected brain is sent to the histopathology laboratory. This gave me newer insights into brain tumours. The fact is most (and may be perhaps all) of them are more or less defined by the imaging features because the brain tissue is difficult to obtain for biopsy. Only in post operative specimens and in cadavers, there is an opportunity to microscopically examine the tissue. And especially in Medulloblastoma, the histology is highly variable and unpredictable, gives hardly any idea about the origin of the disease; the fact is the tumour is not composed of medulloblasts (the type of cells from which the tumour derives its name). The terminology is actually a misnomer and the name is retained only for historic reasons. I thought I should end my research there; this was only getting more complex! I now knew even more certainly that everything would now happen dictated by human behaviour rather than scientific evidence. And I was back again thinking about the people involved; the girl, the boy and their families.

The most I could know was about the girl's family. For them the best prospects were that their daughter gets fully cured and is married at some later date. But how long was the wait going to be and what if she worsens? Should they press for continuing the engagement till then? Or may be find some other suitable boy later on? This seemed even remote because of the unpredictable nature of her illness and the stigma of brain tumour. Too long a wait will get the girl too old to be easily married even if she is cured; we know how particular we are in India about the age of marriage. One can get married even before one gets into their teens (childhood marriages are still not uncommon) but the problem is with aging, especially girls.

The boy had hung around till she was discharged from the hospital and kept calling her thereafter. This must have been tough times. His family was now not too keen on his continuing the relationship with the girl but it would look inhuman if he turned away right then. If not for love, may be for humanity, things had to remain status quo for some time. I also came to know from my cousin about the intimacies they shared during better times (which I would appreciate had continued even in these trying times).

"They had a long period of courtship", she told me. "They would often meet at short intervals, the boy used to come down and they spent a lot of time together. In fact, this relationship was like a lifeline for him. You may not know but his first engagement to one another girl had failed. This had really broken him down but he has sprung back from the crisis. You only need to see them together to sense the romance." God! These were newer insights. Ladylove was not going easy with the boy (It anyway is not easy with most people!). May be I could not sense the element of romance but I sure could sense the tussle going on inside the boy's mind. He almost must have been hoping to turn the tide and rescue the girl from the trap she was in. He must have been now even more smitten by the feelings of love.

What about the girl? She was battling with herself, with the tumour that she was carrying inside her body and also with the changing times. What was she expecting the boy to do? She needed him even more badly but the love must have now looked as sympathy. Neither did she have the courage to ask him to leave, nor could she get in the state to continue. Things would now be moulded by time; time is a great healer; it may not heal her disease but passing time would surely dignify her relationship with the boy. Only with passing time, things could be looked at from a new perspective.

Life went on. A few months later, the girl was not doing any better. She withstood the chemo and radiotherapy well but was now found to have spinal metastasis. This made walking difficult for her, the lower limb muscle power was so reduced that she now had to be on a wheelchair. For the first time she looked diseased. Unlike other tumours which have a remarkable catabolic effect on the body, leading to weight loss and emaciation, the body remains well preserved with brain tumours. Their main problem is due to their behaviour as space occupying lesion, leading to various neurological deficits. This had now started showing. The girl was now feeling even more dependent and perhaps it was this that actually got her to take a decision. She could not pass on the effects of her disease to her loved one. She now wanted to bear the sufferings in privacy. It was only the love for the boy which made her to decide that the relationship could not continue. She politely refused to carry on the relationship with the boy. The disease had found more than one victim. And although the girl survived, the relationship did not.

I could not stop myself from thinking about the boy. Heart break once and then again. If the girl were his wife, there was not much to decide on, he would have stood by her till the very last. Even here he wished to do the same, but was rightfully denied by the girl. If she had lost her life due to the disease, the relationship would have ended anyway, but she was alive and there was a need to duly terminate it. This must have been painful. The girl allowed him to have another go at life but at the same time was ready herself to suffer alone. The boy lingered on for more than a year after they had parted but thereafter he was married to someone else.

It has been a decade now. The girl is still alive. Life is difficult for her but I don't know what troubles her more, the physical disability or the mental agony. This was a lot for her to bear. But she went through it all. Although she could not win over the disease, she must have won over the heart of the boy, even if it meant calling it quits.

TAKEN FOR GRANTED

You should not take things for granted in medicine, we are repeatedly taught this and I have seen it to be true on so many occasions. But that was about disease and its treatment; in our behaviour too, we always make some presumptions. That's part of our human behaviour, and in our interaction with patients, we not only exercise our medical knowhow but also our basic human instincts. These instincts can sometimes get us into uncanny situations; sometimes they can also put us in trouble! I too went through a series of such episodes, where my presumptions put me in a spot of bother!

It was my days as intern in my medical college where I was posted in the department of Gynaecology. The huge load of patients gave us an opportunity to see patients independently and I was particularly excited about that. We had a special cell for contraception counselling and I enjoyed the clinic because it required limited knowledge, the methods were well defined and our main job was making the communication to the patient. Something which I thought I was really good at! But the lesson learnt was otherwise! I had this young couple who were recently married and were looking for a contraceptive solution. They had come from a remote village and it was good to see that the idea of contraception had reached that far and deep. I wanted to live up to their expectation.

"I need a solution which is safe and 100% effective," said the husband in a domineering tone. Everybody desires that, I thought.

"The best for you would be contraceptive pills; they are 100% effective and are ideally suited to a young couple", I said.

"But pills would have their side effects?" he seemed concerned and a little less convinced.

"Not so with these new low dose pills, they have been formulated after a lot of research and found to be almost 100% safe." People are always fascinated about newer things and I thought this may just be enough to convince my patient.

"I take your word. But it must be effective," he reasserted, almost with the look that he would rip me apart if it was otherwise. Or maybe he wanted to show off his might in front of his newly married wife. The Indian male always does it on the excess side when they are with their spouses; probably they want to justify their own domineering attitude with their spouses.

I just gave a smile. You never know about the uncertainties! And in medicine, you generally don't give a word!

"How is it to be taken?" he asked.

"21 days at a stretch, once a day, starting from the 5th day of the menstrual cycle," I said, showing him a strip of the contraceptive pills.

"Thank you," for the first time he said words of gratitude with an equally befitting expression.

They left my room. I realised I hadn't had a word even once with the lady. In fact, there is some reluctance both on part of the patient and the doctor to discuss these issues with women. It was still the India of the late nineties, although not much has changed till now! And with men around, sometimes even the woman's general day to day health care issues are discussed by the male counterpart. Anyway, I was happy that the interaction went off swift and successful; I could convince them very quickly!

Around one and a half month later, he came back. This time he was alone. He came to the clinic and appeared furious. "Do you remember having seen me last month, you had promised everything would be 100% but it was not to be. My wife is pregnant." I couldn't make out if he came to settle scores or if he wanted to terminate the pregnancy. In India you can get a legal abortion if the pregnancy is a result of a failed contraceptive method. But then his wife was not around.

I was perturbed. Oral contraceptives fail only if they are not taken as per schedule, or if they are taken with medicines which can decrease their efficacy. I needed to find out what went wrong. "Were the medicines taken as per schedule?"

"Yes." His voice was now even louder. It was as if he was not going to take too many questions and I had to identify the cause quickly.

"Were some other medicines taken as well?"

"Nothing. And you told me that the pills would have no side effects but I ended up having nausea every time I took them."

"But why did you have nausea?"

"Because I took them," he was simmering.

"I mean but why did you take them?"

"You had only directed to take the pills for contraception."

"It was supposed to be your wife who had to take the pills," I was frustrated and seemed to match my patient in the decibels.

"But you never said that!"

I never did. In fact I did not have even a word with the lady. It was presumed that the pills would be taken by the female, everyone supposedly knows that! But not my patient in this case!

"And you never asked," I said, match for match.

There wasn't much left to be discussed. My patient realised that it was an inadvertent error. He still would have indulged in a blame sharing exercise, but was silenced by my reaction which was equally imposing.

He left the room. It would have been extended frustration for both us had he stayed back. He must have been still frustrated but I couldn't help laughing, probably at myself. I should have understood the man's psyche. In India fertility is all about male factors and infertility is always due to problems with the female, which scientifically is not true. But cultural values and beliefs have stronger roots than scientific facts; I should have considered that, especially as I knew he was from a remote village. It was a lesson learnt, but there were more to be learnt.

This time I had matured, it was more than six years after my post graduation and all my skills had now become refined; but instincts and reflexes are difficult to hone. This was an urban setting and my patient was a lady in her late twenties, accompanied by a gentleman who appeared of similar age.

"How can I help you?" I started.

"I have been having urinary burning on and off for quite some time now and also I have foul vaginal discharge in between," she started off, straight to the issue with clear candid description.

"Is the pain related to the menstrual cycle? It is worse during some particular time?"

"My menses cause some discomfort but that has been the same for years now. This pain is unlike that. I have noticed that the pain is particularly felt at the end of the first week after menses."

That was a significant description; the pain with the timing was highly suggestive of PID (Pelvic Inflammatory Disease). This may be due to organisms which are sexually transmitted.

"Does the discharge smell foul? Is it profuse?"

"Haven't noticed that, but it does soil my undergarments."

It was an excellent description. The symptoms were highly suggestive of an STD (Sexually Transmitted Diseases). The one thing it definitely requires is a physical relationship. It feels awkward to ask an unmarried girl this leading question; pre marital sex is not uncommon but talking about it is still taboo. And it is very uncomfortable to ask that to a young lady whom you are seeing for the first time. I was saved of this ordeal because the lady here was married and I didn't have to ask her; the vermillion told it all.

It was time to write down the prescription. STDs are generally empirically treated with a set of antibiotics and to prevent recurrence, the other partner also has to be simultaneously treated. This can be more than one, but you do not talk about this when they are accompanied by a male counterpart.

I quickly jotted down the prescription. It was time to explain. "These are two antibiotics which are to be taken as per schedule. If the symptoms persist, I may have to evaluate further. I turned my attention now to the male counterpart. He had been silent all throughout the conversation and I thought that the cultural beliefs have evolved over time; this was an era of women empowerment.

"You will also have to take the medicines in a similar fashion," I told him. He did not react, but appeared to be a little curious which I thought was understandable. How can someone be treated for someone else's problem. I was expecting a query and it came.

"Why do I need to take the medicine?" he asked.

"Actually these infections can be transmitted from husband to wife due to their sexual route of transmission. Thus both have to be treated simultaneously for eradication of infection and prevention of recurrence.

He burst into laughter, so did the lady. I was found gaping. I did not realise what had evoked that response.

"He is not my husband," the lady said. "Actually we work in the same office and he just came along because I am new to this place." Both of them were still smiling.

"Anyway, thank you very much. I will let my husband know this." I did not understand if she was talking about the prescription or the incident!

I could not imagine that the woman could talk with such ease in the presence of a male friend. Generally even if someone accompanies to the clinic, they do not enter the consultation chamber when such issues are to be discussed. Not even for once I thought of asking their relationship; it sometimes actually looks awkward. But what I ended up was even more so! The Indian woman had evolved more than I had thought; the gender wall had started to dissolve. It seemed that finally human beings had become a single species! Now I invariably ask for an introduction if there is a need to know the relationship.

But I erred again, this time it was a different situation, but the embarrassment was no less! I was in the ICU and I had been seeing a woman who had been admitted in the morning and was very sick. I had been told by my colleague that she had been transferred from another hospital and the family carries a lot of expectation. "I think they need to know the realistic; please talk to them during visiting hours. I had met the son and gave a briefing, but she continues to deteriorate and I think there is very little chance that she makes it," my colleague said.

"OK. That will be done," I said with a sense of responsibility. Counselling of family is as critical as managing a critically ill patient. Even if we are staring at death, we need to make it dignified. The family also has to be prepared mentally for the inevitable. But we just can't talk only of death; we are there to give life to our patients and hope to their family. And sometimes in prolonged illness there is an issue of termination of care which also has to be handled delicately. Communication in ICU is a great balancing act. We also need to ensure everyone talks in unison; the treating team, the experts, the Intensivist, the nursing staff; there are so many people who are involved in communication.

I did a quick assessment of the patient. The medical history and the progress notes made by my colleague were of great help. She had long standing diabetes and kidney dysfunction and severe cardiac dysfunction, had recently suffered a stroke and was bedbound for the last four months. Most of this time of her was spent in hospital and now she was admitted with pneumonia. Her pneumonia had progressed to respiratory failure and she was on mechanical ventilator, she required medicines to support her blood pressure and had lately been requiring regular dialysis. She had been shifted to our hospital because we have a strong nephrology unit but what she perhaps needed was the hands of the Almighty. In such patients, it's not about what would happen but it's about when; and in between we wait for a miracle to happen.

It was visiting hour and time to talk to the family. A gentleman who appeared to be in his late fifties came to know about the patient. "I had been here in the morning and was told things are very critical, can you tell me the progress?" he enquired. He seemed very disturbed and did not appear ready for the worst. I had to be very articulate in giving the prognosis.

"I have seen her reports. The pneumonia was associated with a blood borne infection which has spread to the entire body and has lead to multi organ failure. She already had multiple ailments from before which have been worsened further by the current insult. We have started her on antibiotics and she is also undergoing dialysis. In fact every system of her requires support. For now do not expect any good news, the turnaround if it occurs is going to take a lot of time; if she deteriorates further, we will make a communication." I knew any further deterioration actually meant death, she could not have been worse!

"Will she make it? What chances you give her?" He seemed desperate.

"The greatest hope comes from the fact that she is still surviving. Our efforts would take some time but for anything to work, she must be alive. I do not know how long we can carry on." Inside me I knew we were fighting a lost battle, it was only a matter of time. The person still hung around and I thought I needed to pass a personal note. The man appeared to be in his late fifties and thus the lady would be somewhere in her late seventies, I guessed. I had not noted her age in the case record. And considering her age and co morbidities, death should come as a usual consequence. It was better to talk in these terms, that life would be more of a burden for her, a suffering, and death would be a reliever.

"She has been fighting terminal diseases and even if she comes out of this crisis, she would not have a great quality of life. I believe she has reached a time when her age is also her enemy. She is requiring on and off hospitalisation, every insult is worsening organ function further and with aging, organ recovery can be slow and incomplete. There is a time when we need to change our focus from mere survival to meaningful survival. That looks remote. She has lived a long healthy life and maybe we should take things now as a natural progression of events. If you ask me to give an honest and personal view, I do not think your mother is going to walk out of the ICU."

"Actually she is my wife," the gentleman said softly.

I was stunned to silence. I should have checked, if not the relation, at least the age of my patient. Because it was the son in the morning, it stayed back on my mind and I made all assumptions based on that. And everything that I had said was reverberating in my mind. My overestimation of the age of my patient and the consequences I drew from it had now created a new set of circumstances, apart from the acute illness for which she was admitted (I hope the husband would forgive me for that, she had perhaps wilted due to the burden of the diseases). I wished I was given a Code Blue call (it's an emergency we have to attend to in a flash) to take a parting from the patient's husband, I couldn't visualise another way to end the conversation. God listened, the code was raised, the alarm rang and I rushed away. And still all that I had said and done was on my mind. And I knew this habit (or reflex) would die only with me!

DIFFERENT DYNAMICS

The loss of a life is always associated with bereavement of some form or the other. And I have seen different reactions according to the circumstances. Whether it's a sudden unexpected death, death of a young man or a middle aged, death at a ripe old age, the emotions pour from different quarters differently. It could be a son, a wife, a father, a grandson, a friend, a guardian on the losing side and everybody has different reasons for being sad. And generally it concerns how they would miss the person; evokes sweet bitter memories of the bygone and we see this happening often even before the inevitable happens. Being in the ICU, you often speak of the ominous prognosis and do get caught in this whirlwind of sadness, have to go beyond the treatment process, console the aggrieved and for sometime become a part of their bereavement. This is transient, but there was one death which shook me from within, not because there was a life lost, but the way it changed the dynamics of the living ones.

This happened during my training in the medical school. Ours was almost 40yrs old and to me it seemed it had failed to grow. I had some realisation then but realised it fully as I started working in a private corporate hospital. India has an amazing parallel heath care system at all levels; one which is supposedly almost free run by the government and the other which is managed by individuals and corporate which is all paid. And you really get what you pay for; I may appear complaining but the public health care system in the country really needs a relook. It was here, in our government medical college hospital, a gentleman in his late forties was admitted with a big stroke. He had fallen unconscious suddenly at home and his brain scan done in our hospital revealed a large intra cerebral haemorrhage. He was admitted in the general ward which perhaps was not appropriate but at that point of time I had no idea how an Intensive Care Unit worked. We had an ICU in our hospital which had recently started and was infantile, both in its working and its bed capacity. In fact he was fortunate to have got himself a bed in the general ward, for whatever services we provided, we were always full. The man was accompanied by a lady and two small children. The lady seemed to be lost in the cacophony of the hospital, she just did not know whom to meet and what to expect. I was on duty and I could read her face very well. I had both the disappointment and the guilt that there was not much that we could do for the patient, so I thought I should extend some help to lady, who I guessed was his wife. She was at the bedside when I began.

"How are you related to him?"

"I am his wife."

"Actually your husband has suffered a massive intra cerebral bleed and his condition is very critical. There is not much that can be done at this stage and the first few days would be of greatest concern." We were not used to explaining in detail the patient's condition, not only because we were busy but also because we had not learnt it and were not used to it. (I think I should better use the word "I" to personalise, we would imply a lot of people who may not be appreciative of what I say). I was surprised because she hardly had anything to ask. She was as still as the patient. We were done with the other patients and I came back to her because somehow I was not feeling easy. I had actually come back again to start a communication but she still did not talk. I pretended to examine the patient when suddenly the lady broke into tears.

"Sir, I have two small children and there is no one else in the family. They are still school going and my husband is the bread winner in the family. He works in a nearby factory and we somehow make both ends meet. Please take care of him. Please."

I was taken aback by the response. I was expecting her to ask something about the disease, some enquiry so to what was the exact condition and what was being done. Rather what concerned her was the socioeconomic condition of her own, it felt as if she was the one who was staring at disaster than her husband. I could not say much. It never translated into a conversation. It was over for that day.

I reached early the next morning and it was still the lady who was with him. It is usual to have a male guardian around and we generally do all the talking to the males, especially when it's an issue of life and death. You can be more straightforward and blunt with males than with the females and also most of the decision making is done by the males. So much so that we seldom get to talk to the wife even if she is around. Her being alone was only an endorsement of what she had said about her social support system. It's not unusual for women to bring their husband to the hospital but by the next day, most family men are around. She was not from a far off place and I was expecting more people to pour in, also considering that he was very sick. That was not to be. The prognosis was repeated the next day with the same monotony as it was done the other day; short and emotionless.

I had more interests now in the behaviour of the women than that of the disease. I was wondering how she would take the fate of her husband and how it was going to change her own. I was now more observant of her behaviour. What caught my eyes was something unusual. She looked more concerned about her children than her husband, perhaps she knew that she could not do much about him but there was a lot to be done for them. When food was served, she took it in a container but her husband was not in a position to eat. She actually took it and fed her children and did not eat anything herself. It was a mother at work and the one who was now terribly alone. I could not understand if she had resigned to her fate or was actually trying to find a way out.

I had my lunch and was back again. This time the lady spoke.

"Is he not going to make any recovery? Isn't that anything that you can do? We have learnt about this hospital in our village that there are very big and learned doctors here and they can treat any disease. He was alright two days back. What could have had happened to him that he doesn't even respond?"

I realised she did not understand the exact nature of the disease and it would be a difficult task as well. She had always seen him work hard the whole day and that needed a lot of physical strength. Suddenly something had struck him like lightning and he had stopped responding. The fact is the patient must have had high blood pressure because our initial recordings were very high. Also the area of the bleed in the brain was classical of a hypertensive bleed. He was unfortunate at both the ends, lack of good accessible primary health care meant he never knew that he had high blood pressure and now when he needed the best tertiary health care, he could not get one for himself. I realised health had fallen victim to our socio economic structure and even though we know that preventive health care is the most cost effective, our system has done very little to put it in first place. Lack of health care facilities, qualified and competent doctors, free availability of diagnostic services, medicines; we have failed at all places. Even urban India is no good, but the scene in rural India is even worse. All we have is policies and paper work, and promises and hopes built in the electoral season. I was almost losing my focus from the patient and getting carried away in this tornado of thought process. I had to respond to her.

"Actually there is a big clot in his brain and this has resulted in a total dysfunction of the brain function. I am afraid but there is no way the function can be restored except if he can carry through this period and remain alive." I knew I had put it in the worst possible way, almost leaving it to the patient to recover on his own, but that was the truth, at least in the set up and the circumstances we were working. As expected, there was no further query. She must have turned her attention to God and must have been wondering why she was anyway in the hospital. The only thing we did very well was expressing our helplessness and the prognosis of the disease.

The patient died the third day. The lady rushed to the nurse as her husband had stopped breathing. It was my duty to examine the patient which I did. I also had to make the death declaration but I did not have the nerve to do so. I asked the nurse to help, I excused myself by saying that there was only a lady attendant. But inside me was a sense of shame. I expected a loud outburst but it was not to be. I just tried to have a look and what I saw caught me dumbfounded. The lady was clinging on to her children, I do not know if it was to support them or to support herself. The children were old enough to realise the meaning of death but its true implications were known only to their mother. The circumstances were such that she could not even have an emotional outburst at the loss of her beloved. Her miseries were even more; life ahead seemed even more difficult than what she was in already.

I could not share her sorrow but was deeply affected. It was life lost not only to a disease but also to a failed system. It was the death of not just an individual but a family. The poor social support left the lady in such a state that it denied her even the bereavement of the most beloved. I knew she had to come to terms to the changed dynamics, without a clue, except for the surety that life would never be the same again.

TO EAT OR NOT TO EAT

A consultation with a nutritional expert is standard in modern medicine for most chronic diseases but nutrition as a therapeutic module is not standard. It is always considered as add on; drugs are still at the centre stage in disease management. I also have my doubts as to how many diseases can be treated only by nutritional therapy; there is hardly any data except for obesity, malnutrition and nutrient deficiency disorders. Our knowledge about nutrition is also poor; the number of pages dedicated to food and nutrition in our textbooks is minimal. The most we read about is the complex biochemical reactions and pathways which have no clinical application. We read about carbohydrates, proteins, fats, vitamins and minerals, but what people eat is food and this translation from molecular biochemistry to kitchen science makes a mess of all the theories and formulas we know of! Add to that the variation of food habits among people; implementing a dietary plan can be difficult both for the patient and the doctor.

The way a nutritional plan is explained and understood relegates its importance even further. The focus is usually on the delineation of food one should not eat in a particular disease rather than emphasising on what one should eat. Consider this scene in a patient with elderly age, multiple diseases and poor appetite; it is sure to spell doom. Also, the domain of nutritional therapy generally rests with certified nutritionist who have little knowledge about the disease process and the doctor who looks after the disease process have little idea (or better still interest) about the nutritional issues. Unlike the regular follow up the patient has with their doctor, the consultation with a nutritionist is generally a one time affair. So nutrition also loses the importance it should receive as a therapeutic tool. This was best exemplified by one patient I had to see for weight loss and it made the experience even more remarkable because the patient in this case was himself a doctor!

I was in the hospital when I received a call. "Sir, there is a referral for you, the patient is in room number 108," the nurse said. It is not usual for a young general physician like me working in a super speciality hospital to receive a referral; I thought it must be some patient whom I could have seen earlier and my services would be merely be of token nature. I got into the room and saw an elderly gentleman whom I did not recall having seen before.

"Hello, I am Dr. Rakesh". I did not want to qualify further that I was the Intensivist at the hospital; this could have made the patient feel a little scary, no patient sitting comfortably in his room would appreciate that someone from the Intensive care has been asked to see him.

"Hello, please have a seat," the patient said. I could make out that neither the patient had seen me before nor had he ever heard of me.

"I am a medicine specialist and have been asked to have a look at you by your treating team. I have gone through your case record and I have noticed that you have lost more than twenty kilos of weight in the last two years. I also came to know that you are having diabetes for the last twenty years and that you underwent a coronary bypass surgery three years back. Of late, you have also developed kidney dysfunction".

There was nothing to smile about all that I had said; for most patients there isn't much to smile about when they are undergoing treatment in a hospital. I had started on a very sombre note and needed to say something to lift up the spirits. "I also learnt that you are a doctor yourself and this came as a pleasant surprise, it almost feels one generation is looking after another." My eyes were wide open and my every facial muscle was at work to create a dignified smile and it seemed to work.

"Yes, I am also a general physician working for more than 35 years now. I must have been as energetic as you are but these few years things have been a little difficult." That was obvious. Even if I did not have a look at his medical records, I could have made out that he had been keeping sick. His hair was coarse and dry and so was his skin and his facial bones were easily appreciable even though he had been keeping a thick beard. The wasting of his arm muscles was visible as he was wearing a vest and the scapular bones almost stood out like wings on his body. His thigh and his legs seemed almost equal in girth, but his belly was still full; this was the classic picture of muscle wasting, the lemon on tooth picks appearance.

The load of his disease was enough to make him feel lethargic, which was worsened by the unabated weight loss. He looked worse than what the lab reports suggested. We do see emaciation in chronic organ failure, whether it be kidney, heart, lung; but it was too much in too little a time and he had not reached any terminal or end stage organ dysfunction.

"Have any of your previous work up revealed any reason for your weight loss?" I enquired, because he had been under constant medical follow up and had a bulky file of various investigations.

"Nothing specific, but I fear there might be some mitosis." That was an excellent choice of word for cancer, considering that his wife was around and he did not want to send further shockers. In the elderly cancer is a strong consideration for unexplained weight loss.

"That would be clear from the imaging results and the biochemical tests that has been done, although none is suggestive as yet." I wanted to be re assuring and also wanted to change the focus. Was the most obvious thing being missed, poor intake leading to protein calorie malnutrition? It was time I took up the issue because this may not be so easily reflected in the tests, it can only be ascertained by enquiry.

"How is your appetite these days?" I continued.

"It has been poor. I just do not feel like eating."

"Is it because of the kind of food that you are given or because you genuinely do not feel like eating?" I was almost certain that a lot of restrictions must be in place.

In the presence of multiple diseases, the appetite anyway is low and this can be worsened by selective and bland food.

He raised his eyebrows and looked at his wife, needless to say that he was unhappy the way he was being fed. The average Indian wife is very protective; they treat their diseased husband like small children and I knew his wife wouldn't be behaving any different. She must have felt it her solemn duty to give him (or not to give him!) whatever the dietician must have told (I am now switching over from the term nutritionist to dietician, because this is what they end up with!). I knew it was time to talk to the wife.

"Do you think he is eating adequately?"

"There is so little that I can give him in the presence of the diseases that he has," she did not want to answer my question but I understood what she meant. It was good that my patient was a doctor and I could discuss things in technical language; otherwise discussion on food can be long and messy.

"I can understand your problem. The dietician would have restricted your carbohydrates because of diabetes, you must be on restricted proteins because of kidney disease and fats would also be restricted because of cardiac disease," I looked at the patient and said. "And now that your potassium levels are also high, the choices must be even more limited. Did you ever consider how your daily calorie requirements are met?"

There was a pause. And then there was a sense of defiance. "You need to talk that to my wife." I knew I had to. All the three forms of food which provide energy were restricted. The only other form of food we need is water which is truly a zero calorie food. From the diet chart that was prepared for him, I also came to know that he was a Vaishnavite and had adopted vegetarianism. That made the options even smaller. Nevertheless, there is a trend among patients to turn vegetarian the moment they are declared to have major diseases or organ dysfunction. There is a strong cultural belief that eating non vegetarian food is not good for health, especially when someone is sick. In fact, people give up spices, condiments, change the way meal is prepared; bland and boiled food is supposed to be of therapeutic benefit. People often make the disclosure about this change during interviews; even the most anorectic patient is seen to follow this policy. There is no scientific basis for all these; in fact it worsens their nutritional status even further!

I turned my attention to the wife. "He should be taking at least 1500cal a day and should be allowed at least 35gm of protein daily. A litre of whole milk would provide about 750cal and about 18gm of protein. The pulses on an average provide about 20gm of protein for every 100gm and the grains give you half of that for the same amount. Fruits and vegetables have negligible protein content. The fact is that a vegetarian even if allowed eating unrestricted can hardly procure the 60 to 70 gm of the daily recommended protein intake. For him this is reduced by only 20% because of kidney dysfunction and in extra ordinary circumstances by at the most 40% and even then it will be difficult to provide him that protein amount." I had the statistics ready with me because of my interests in obesity management but here they were to be useful because of different reasons. "And if he is not allowed adequate calories, the body starts eating on its own muscles for energy and that is what has caused a loss in his muscle bulk."

"He is not able to eat anywhere near that," the wife said. "He keeps asking for his choicest food but I have been unfair", she was feeling almost guilty.

"Give him energy dense food. Fats would be the best in that category. Sweets, chocolates, ice-cream, dry fruits, pea-nuts, fried potatoes, sweet yoghurt etc. there is a lot to choose from." I was particular to mention some foods and recipes to drive the point home.

"God, I might feel charged up if I am allowed to eat this freely," my patient brightened up. In the corner, the wife was pleased to see her husband's face glow, she must have been missing that look for a long time now.

"Also give him carbohydrates unrestricted, be it simple sugars, grains, fruits, bakery products etc." I said. I was adding every class of food to his menu.

"But what about his blood sugar levels?"

"We do not control it by dietary restrictions at the expense of inadequate calories. As a vegetarian, anyway you would be on a carbohydrate predominant food, be it grains, simple sugars, fruits and vegetables; they are all carbohydrate rich. If your sugars start rising, we can control it by insulin. That would be a better bargain and considering that insulin is an anabolic hormone, it might help you build up your muscles." It always makes sense to explain things scientifically; otherwise it appears to be a personal and biased opinion. If someone else disagrees with your opinion, the patient has some scientific basis as to whom and what to follow.

"In fact, sometimes the essence of these dietary recommendations is lost; they are meant for people with good appetite, ideal body weight and proper body proportions. The problem with you was that first you had dietary consultation for diabetes, then for cardiac disease and then for kidney disease and all of them must have been in isolation. And there must not have been any review."

"I am surprised I did not have this thought cross my mind. The complexities of modern health care have made some of these basic issues take a back seat. I should have realised this as a doctor. I am thankful that you brought up the issue."

"That is because I knew everything else would be taken care of," I said with a smile, which was reciprocated with even more warmth. "And for some time, forget what the dietician said and make your own food plan, with maximum emphasis on adequate calories no matter where it comes from. But it surely should include all that your taste buds yearn for. And may be ask me to join for a sumptuous lunch tomorrow!" we all cracked into a laughter in unison. It felt like this had been missing for a long time.

"I will take your leave now. We will meet tomorrow." I left the room, which was now filled with a sense of hope. I do not know if his physical condition could be attributable only to malnutrition but what the weight loss had done had not only made him physically weak but also mentally depressed. I could not have done much for his co-morbidities; he had already been seen by the best in the field. And my patient was also a doctor who had a lot of insight into his disease, for him it would have been very disappointing that his health had been deteriorating consistently. It can be ascertained from the fact that the reason he got admitted in the hospital was his weight loss. My job perhaps was to give him a ray of hope, a straw that he can cling to and I was happy that I was able to do that. And may be on occasions when he eats like a king, he remembers me with a big smile.

DISEASE REDEFINED

According to WHO, health is a state of complete physical, mental and social well being and not merely the absence of disease or infirmity. If this is to be followed, probably most of us (or may be all) would not be considered healthy most of the time. If we look at the word disease, it comes from dis-ease, which means not at ease; essentially pointing at the symptomatic nature of diseases. But medicine has evolved from symptom based health management to identification of factors which can lead to ill health later. Some of the most common diseases like diabetes, hypertension and dyslipidemia fall in this category; in most patients they do not cause any symptoms, they are at best to be considered as risk factors for development of ill health and organ dysfunction in future.

One condition which is of similar nature and perhaps even more prevalent is Obesity, but it is seldom considered as a disease. One of the biggest challenges I had when I started my obesity clinic was to emphasise obesity as a disease rather than just a cosmetic and numerical derangement. What I experienced changed my concepts about health and disease even further and also influenced the way I dealt with obese patients.

I met a 39yrs old entrepreneur, who had started a chain of educational institutes and had been making professional progress at an alarming pace. His work entailed a lot of travel and stress which was causing health concerns and maybe he was aging faster due to the speed at which he was doing things! He decided it was time to get a preventive health check up (which in India means getting a few set of laboratory tests done) and came to see me with the reports.

He had the typical metabolic look which I have in my mind; his hairline was receding, he had a prominent paunch, dark lips and all the typical curves which were appreciable in his body hugging T shirt. The only thing which looked sporty about him was his sneakers (and I was sure he was not doing any justice to them!).

"I have been diagnosed to have high blood sugar; I had got a health check done," he seemed to be in a hurry for a solution.

He immediately handed me over his reports and I saw his blood sugar was 380mg% (ideally it should be less than 140%). "Do you any other problem?"

"My cholesterol level is also high. And the other day the nurse found that my blood pressure is also high.

"Anything else?"

"No. I don't have any problems. I just realised them because of the tests."

"Do you smoke?" I asked. I knew he did, the smell of nicotine was still fresh; my only interest was in how much.

"Yes, I smoke more than a pack a day; it's almost 15 years now". May be he knew I would be interested in the amount.

"Do you consider your smoking as disease?" I threw this question only to make him consider if he needed any treatment for that.

There was a pause. I had to go on. "What about your weight?"

"I am ninety five kilos". For a person of 168cms, he was almost 25 kilos overweight.

"That is much more than normal. Did you ever think you needed to do something about it?"

His silence told it all.

"So you felt the need to meet me only when you saw the blood reports. Why are you worried about your blood sugar, your cholesterol, your hypertension; they haven't caused any discomfort. You just came to know they are above normal limits, how does it bother you?"

"There is so much said and read about them", he was miffed. "These conditions can lead to so many problems in the future; I want to get things corrected." He really seemed to be concerned and committed.

"Then why don't you consider your obesity as a disease? This also can lead to a lot of problems, not only metabolic but also mechanical." I almost wanted to say you already have many of them and many more will keep coming.

He almost smiled. His mood changed in a flash and he understood why I had been gruelling him till now.

"If you are here to take my opinion, then your only disease is obesity and smoking (or may be your lifestyle! I did not want to use this term because the general perception was that he was successful and I did not want him to change that style at all). There is a general principle; in the elderly, you may have many diseases for a single problem and in the young, we generally look for one disease which explains all problems."

He was impressed with the idea, more so because I considered him young. Although he had been withered because of the load that he was carrying, both of his body and his work, he still had not reached his forties and had to be considered young!

"Everything else that you are having is actually a result of the Metabolic Syndrome which obesity has caused in your body; instead of considering yourself as having diabetes, you should consider yourself as having Insulin Resistance Syndrome. And if you look at the health impact of tobacco and obesity, they are respectively the first and second leading cause of preventable death in the USA". (Somehow people are more convinced about the American statistics, probably also because they know we don't have too many authentic ones of our own).

This sent alarm bells. The fact is this data sounded alarming even to me when I read it first. Actually both have association with cardiac disease and thereby automatically climbed up the charts.

I do not know if this caused more confusion about what disease he had or if it clarified the matter. However, he did not ask me anymore about what his disease was; he was now only interested in the solution.

"What do I need to do now?"

"You need to cut down the Insulin Resistance Syndrome and this is by cutting down weight. It can be achieved by exercise and dietary modification. You probably must have already heard about the importance of food and exercise in diabetes; their contribution to weight loss makes them even more pertinent in your context."

"You mean to say that I can get things normal by just loosing the extra weight?" he seemed both curious and sceptic.

"That's worth a trial". I would have been compelled to start him on a medicine if he had come to me with symptoms, sending a patient back without a medicine in those circumstances would seem unethical and unacceptable even to myself. But he had known the fact only by chance.

"If you promise to follow up, I can give you a month and only then consider the need for medicines. And then perhaps you will also realise the power and importance of these interventions." I wanted to give my weight loss mantra a therapeutic module.

His face brightened up. He was still in his late thirties and must have abhorred the idea of taking medicines for good health; it generally makes you feel even more sick and also old. The non compliance is also greater when they are given to patients who do not have symptoms. I could have prescribed a medicine, in the general circumstances I would have done that, but this was a special situation. If these patients are sent off without medicines, one thing they surely do is follow all the other lifestyle measures because there is nothing else to depend on. Although I would still accept that it is easier to convince people to take a lot of medicines (including insulin) than to get them to exercise and lose weight. And there is an added risk, if they don't lose weight, you are sure to lose your patient!

"I am definitely going to follow what you have asked. Even I want to see how it changes it my metabolic profile."(I was happy he had picked up the idea and also the word!).

"One thing it is definitely going to do is make you look younger," both of us broke into laughter in sync; the sense of humour has to be good to make the statement look inoffensive, I was happy my patient had it. He was only two years elder to me, both of us were also dressed up the same; jeans, T shirt and sneakers, but there appeared to be a generation between us. And not because I looked any younger!

"And this is something you would not able to achieve by a medicine, even if it controlled all your metabolic parameters", I continued, almost emphasising that he got back into shape (and not for cosmetic reasons!) One another thing you will have to do is to cut down or quit tobacco."

"Does that also help in losing weight?" he seemed almost obsessed (or may be possessed!).

"No, it doesn't," I could not help smiling. "But it surely significantly decreases the incidence of cardiovascular events and all my concerns regarding your health are aimed at improving your metabolic profile." I was not sure if he would cut down on smoking; in this world where people are bothered about immediate results, such long term goals are hardly taken with any seriousness and my patient's profile was surely of someone on the fast lane.

"I will give up on that and I will ensure that there is a turnaround very soon", he said. I was not sure if he was making a promise to me or to himself. But the treatment process had one effect; everything was now left to the patient. My role was to show him the way; it was now his turn to tread on it.

He left my room with a new sense of energy, a sense of purpose. It could have been a load of diseases with an equally heavy prescription of medicines, but both of us behaved otherwise! I was still wondering why people forget to appreciate the most obvious, you do not need the know the complex equations of waist circumference, the waist hip ratio, the BMI, the appropriate height for weight etc; you just need to stand before the mirror to know that you are obese! And truly it is this abdominal fat which matters, rather than the overall weight you have. I realised from this patient that the importance of obesity is actually camouflaged by the very diseases that it causes; diabetes, altered cholesterol, hypertension, cardiac disease, osteoarthritis etc. These names are too big as diseases compared to obesity. The behavioural nature of therapy, lack of a pharmacological cure; obesity just does not seem to be the domain of modern medicine! And it is not surprising to see that the management of weight loss has shifted to the cosmetic clinics and people who make claims to weight loss without exercise, diet and lifestyle modifications. There are more obese people ready to take these unconventional measures than try the established, conventional and most effective weight loss measures because there are very few organised clinics providing conventional cure. Even at my clinic I was almost a single man army; the nutritionist, the physical therapist, the doctor, the motivator; I knew I could carry on only because it was early days, few patients and a lot of time! My thoughts were drifting and at the same time I was both excited and wary of what happens to this patient of mine. I was more concerned about the reactions than the results!

And I got the first reaction three weeks later, a week before I was expecting. It was a phone call which my patient made.

"I have lost three kilos of weight and I have also stopped smoking."

"That's great news," I said. I was happy but equally surprised; he started by first telling about his body weight rather than his blood sugar level.

"My exercised capacity is slowly increasing and I am feeling very good about life," it looked like the words from a man who had made a new discovery.

"What about your blood sugar levels?" I actually had to make a query.

"My fasting blood sugar is approaching 200mg%."

That was fantastic, I was telling myself. And even though he could not see me, I was grinning at my broadest! This was more than what I had expected.

"I will see you after three weeks, I will be travelling these days and I remember the tips you gave me to be able to continue my weight loss programme during my travel. Bye"

I was overwhelmed. I had actually discussed with him how to continue exercising and maintaining the diet plan during travel because I had known his profile. It was good to see he acknowledged the same.

We met again. I was seeing him after ten weeks. He definitely looked ten years younger. He was still wearing a body hugging T shirt but the metabolic contours were missing. His blood sugar had normalised and he had lost a full six kilos. He was truly someone on the fast lane; this turnaround was amazing! The meeting was brief, for him it was more of thanksgiving than a check up. I never saw him again. I was bothered I may not see him if it doesn't work; I never knew he will not have to see me because it worked so well! I had tried to redefine his disease; and he ended up redefining his health.

THE DENGUE EPIDEMIC

It was the first time I saw the outbreak of Dengue in the local community. Although it was the first time there was widespread incidence of Dengue, the name was known to almost everyone, all because of wrong reasons. A single death due to Dengue is enough to hit the national headlines, the media more than eager to mount an attack on the failures of the government but the actual impact it leaves on the psyche of the people is altogether different! To him it does not mean failure of the public health care system, it actually has more personal implications. Dengue to the average Indian spells dread, it almost feels once infected you are doomed unless there is a heavenly intervention. The public reaction did not come as a surprise to me, but I thought all this can be changed once you counselled the patient. I couldn't have been any more wrong. What I went through was a lesson in learning as to how the public perception about a disease affects every aspect of disease management and however hard you try, it isn't easy to change the mindset.

This was around summer when cases started pouring in and they were everywhere, in the outpatient department (OPD), in the general wards and in the intensive care units, reflective of the large spectrum of the disease. Before I narrate my experiences, let us share a few facts about Dengue. Dengue is an infection caused by a virus and carried from person to person by the bite of a particular mosquito, very similar to what is seen in malaria, except that malaria is caused by a parasite. In most cases, the infection causes fever, severe bodyache and other constitutional symptoms, frequently goes unnoticed, either because the patients takes some over the counter medicines or because the doctor doesn't feel the need to do any blood test on such patients. Considering that the test itself costs around a thousand rupees and medical care in India is not so comprehensive, this is the larger scenario. In a select few, difficult to quantitate what percentage because of both under detection and under reporting, the disease may get complicated, the patient may develop generalised bleeding, shock, multi organ failure and may succumb to his illness. So it could be an innocuous fever to a life threatening disease and you never know which way the disease would move. Although there are predictive models and warning signals about disease severity, they are not without fallacies. To make matters worse, there is no specific treatment for Dengue. It's all supportive and symptom based. Analgesics, antipyretics and a lot of reassurance (the latter perhaps being the most important one) for the uncomplicated ones and blood products, IV fluids, life support or organ support systems for the critically ill ones. With appropriate and timely treatment the critically ill ones mostly can be saved but I have seen mortality despite the best efforts, underscoring the lethal nature of the disease. Morbidity, mortality and the health care burdens all of which could have been avoided by public health care measures. You may not have the antiviral for the Dengue virus but you have the armamentarium against the carrier mosquito and that is where the entire preventive care strategy is targeted at.

Thus any epidemic of Dengue implies a failure of the public health care and sanitation systems, the primitive and basic issues which we still haven't been able to achieve so many years after independence. And every time there is an epidemic, there is a lot of media attention, statements by bureaucrats and politicians explaining the stop gap arrangements they have made, advertisement splashed all around on measures of personal safety (left to the individual to take care of himself. Sometimes it isn't unusual to find a hoarding right above a garbage dump!), regular alerts and so much more. This creates the effect that a devil is around and anyone might be the next victim. The fear is all too obvious once the diagnosis of Dengue is declared to an individual, the all too nearly normal person stops feeling so even after repeated reassurance, the one who was feeling sick in his body now feels sick in his mind; anxious, fearful, nauseated, filled with a sense of impending doom.

These create problems of a practical nature both for the patient and the doctor, both of whom may be apprehensive, the doctor trying too hard not to be on the wrong side of things and the patient too scared even when things are going right! So you may have a patient with uncomplicated Dengue, due to recover uneventfully over a few days but who just cannot come to terms with the illness. And as a doctor I need to participate in their gloom, counsel about a flu like illness as if you counsel about a critically ill patient, and even if everything is alright alert them about the warning signals which make the whole ambience even more sick. My sense of humour just doesn't help, because the scene is too tense for it, I might be rendered insensitive and careless if I ever did so! What does all this do to me? Makes me feel even sicker than the patient!

One of the best ways to avoid this situation is to skip the test for Dengue because it hardly makes any change to the management strategy in the not too sick patient. This may work well in a subset of patients who have limited resources but I realised that wasn't the right way because it is my job to reach at the right diagnosis and initiate the appropriate treatment accordingly, conducive to the resources of the patient and his needs. The other way is to hospitalise all patients diagnosed with Dengue to give them that false sense of reassurance which they so badly need by putting them in a health care facility. This saves you from the unnecessary phone call either from the patient or their known ones (which in India is a very prevalent method of patient doctor interaction), unnecessary follow ups and also does a little good (or may be a lot good) to your coffers; also alleviates you of all the worries of the uncertainties. Thus anything going wrong would be blamed on the God and not on you! But I couldn't reconcile to this method. I thought I would stick to the basic way and treat my patient the way he should be treated, copy book standards. And probably it was this that created the experiences.

My first few patients went off well. They all were from the average middle class, average IQ, average resources and had an average disease. My standard explanation to them was, "You are very lucky that your Dengue hasn't become complicated and if God is by your side (it's always good to form this team; you, patient and the God), you will make an uneventful recovery. All you have to do is taking adequate rest, good food and avoid unnecessary medicines." And they actually start feeling lucky. God has been so kind to them that their dreaded affliction is blissfully without any significant harm. And the moment they feel a little bad, their mind turns to God rather than their doctor. So far so good. Wasn't to be. This time it was an upper class middle aged man with average grey matter but more than average in his coffers who was diagnosed to have Dengue on serological testing with three days of fever and body ache. I had the fortune to see him before so a rapport was struck as soon as we met again. He came along with his wife who appeared more sick than the patient and most of my time was spent in treating (read communicating) her. After a thorough examination and discussion, a consensus was reached that he can be treated on Out Patient basis and they can get back to me whenever they feel uncomfortable. "Do I need to do some blood test for regular monitoring?" asked the patient. "Not necessary", was my nonchalant reply.

"Not even platelet count?"

"Only if clinically indicated", I emphasised.

Two days went by. I got a phone call from my patient. "Doctor, I got my platelets checked. Now they are 80 thousand, two days back they were 160 thousand. I am a little concerned". He was only little concerned because his fever had come down and symptomatically he was stable, otherwise there would have been panic.

"Do not worry, you are still in the safe range, platelets should not worry you unless you have some evidence of bleeding or the counts come down to below 20 thousand. Even at those levels you will not necessarily need a transfusion, but may be closer monitoring and a possible hospitalisation."

"Thank you doctor."

It was a good feeling. I was priding at myself at being able to assure my patient. 15min on and another phone rang. "I am the elder brother of the patient who is suffering from Dengue and just had a word with you".

"Yes. I recall. Your brother seems fine and should make an uneventful recovery". I said in a composed tone.

"But his platelets have fallen; I think he needs an urgent hospitalisation". His tone was protesting.

"I know that, but even at these levels there should be no concern". We will see if they fall any further or if he bleeds".

"Do you mean to say we should wait till disaster occurs? We need to do something urgently, I am going to get him hospitalised. Let's see what the Doctor says."

My ego was annihilated. Who was I? And what was I doing? Do I really need to treat patients on demand? I was following what the recommendations said but they have been made based on medical facts and not patient's perceptions. These are best followed in countries where heath care is free and delivered entirely by the government and you are under the compulsion of doing what has been laid down in guidelines. Not so in India. Here you can choose your Doctor; sometimes you can even choose your treatment! The patient bears the bill in the private sector and he wants to have his say. I seemed to have lost to the system.

I was still curious what happened thereafter because there were no more phone calls (expectedly!). A few weeks later, I came to know through another patient of mine that he was hospitalised, received a few units of platelets (which must have been on demand!) and made an uneventful recovery. Someone did better than me I thought. But what was more of concern to me was if I needed to change my approach. Difficult, I thought, the best way was to continue the same way and see how people react, I may lose another patient but that might give me another story! Newtonian theory!

My next patient was someone whom I knew, he was my classmate Dr. Sushil's uncle. We had spent a lot of time together during our graduation years and I knew his whole family. He called to say that his uncle who stays in Arunachal Pradesh has been diagnosed with Dengue and is coming over to Guwahati for further treatment.

"They might reach somewhere in the midnight. Where should they see you?"

"Is he really that sick that I may have to see him in the midnight and hospitalise him?"

"Doesn't look like. He is coming by a private vehicle and not an ambulance and he seemed okay over the phone, except that he was sounding panicky."

"OK. In that case ask him to take some rest after he reaches Guwahati and he can see me in the morning. In case he feels that an urgent consultation is needed, ask him to call after reaching".

There were no phone calls during the night. At seven in the morning, I received a call."Rakesh, its uncle here". "There was only one uncle who could have called me that early in the morning and we came straight to the point.

"Yes, I know you were supposed to reach late night and I was expecting a call. I have also been briefed about your disease." I wanted to be re assuring from the word go.

"I reached here at 5, put up with my cousin, had my bath and breakfast but I think I should get to you in the hospital."

"I will be there in an hour. Come over." It was going to be an early start to the day.

I reached hospital at 8 and he was already there. I passed a courtesy smile but there was no response; expectedly. He quickly got into my OPD clinic.

"What problems are you facing now?" the usual start to every patient.

"I have Dengue, don't you know?" his reaction was as if I was caught unaware.

"Yes, I know that, but currently what are the symptoms?"

He felt bamboozled. What else I needed to say, he must have wondered. The pause made me throw another question. "I meant I know that you have Dengue. I just needed to know when it started, where did you get the tests done, are you feeling worse than before and did you receive any kind of treatment"

My composed behaviour seemed to have an effect and I could see his nerves cooling down. "Well, I just had three days of fever with body pain and I got the test done last evening and as soon as I came to know that the test was positive, I rushed to Guwahati. You know health care facilities there isn't so dependable, so I thought it was better to be here". It seemed quite sensible. Every doctor desires that the patients reach a dependable health care facility, so that in case of complications, there is no precious time lost in initiation of treatment.

It didn't take me long to finish off the interview and the clinical examination. "You don't have any reason to worry; I hope you will make an uneventful recovery. Stay back in Guwahati till you feel fully comfortable and then you can get back to your place. The only medicine my prescription carried was Paracetamol as needed for fever and pain.

"No need to get hospitalised?" was his instant query.

"No. And you won't need any tests as well. As yet, things are fine".

Was it for this that I made a long overnight journey, he must have been feeling. Can't say if he actually felt cheated! Came with the expectation that there was so much to be done by modern medicine and a family physician but the ceremonies ended too soon. Almost the feeling you have after seeing a James Bond movie where the Bond is without his gazettes and gizmos. Very similar to what I felt after seeing the Skyfall. (But I heard it still did well at the Box office).

But the pause this time was broken by the patient himself. "Actually, I reached hospital much early and I rushed to the lab, submitted my blood sample, ordered for a few tests myself (you can do that in India) and told them if something further was needed, I will get back to them after seeing you. I thought this will expedite the process of reporting. Anyway, with whatever that has been done, I will see you in the evening. And yes, I haven't come alone; I also have my friend with me diagnosed with Dengue and who has come along with me all the way. Please have a look at him".

His was also a similar story. What must have been reassuring to him was that both of them were being treated in the same way. After they left, I had some time to think back to myself. If it weren't for my friend's relation, these two would have also left me for another doctor. I wasn't sure if they might still do it. But this time I would get to know it from my friend.

We met again in the evening. This time the atmosphere was light. I could see both of them smiling and we actually broke off with a personal note. "It's good to see you after a long time. I am told you are doing well and that you will take full care of us. I had talked to Sushil after meeting you and he told me to follow what you said. We were very tense last night. We actually came in two different cars, in case there was a breakdown (which is not unusual in that difficult terrain), we will still have a way to reach early without delay."

"I actually came to know about my report being positive late evening", his friend said." I was still in the town with my car and decided to leave immediately; I didn't even take a pair of clothes. I was scared and clueless. And then after discussions your name came up and that's how we have reached you. We are really appreciative of the way you managed us. Thank you very much."

I was feeling relieved and it also restored my sense of pride a little. Till now my curiosity was how patients handled the situation, now I thought it was time to see what the medical community was doing. The scene was grim. You could see patients admitted in hospitals who perhaps need not require institutional care but somewhere the doctor had succumbed to the desires (or fear) of the patient than to the severity of the illness. My curiosity rose further so as to see what treatment was meted out to the patients in the absence of a specific therapy. I decided to visit a few hospitals. Identification of patients with Dengue was easy as they had to stay in their bed with a mosquito net all the time so stop the transmission of the disease. This was a praiseworthy step taken by the health authorities. (I appreciated it more because it allowed easy identification!). Reporting of Dengue was made mandatory (although there are many notifiable diseases this is seldom done unless there is a strict government directive, and when it comes to Dengue, the authorities mean business!), so it was easy for me to collect the statistics. Majority of the patients had uncomplicated Dengue but continued to stay in hospital till their fever subsided fully. And received fluids, antibacterials, vitamins, acid suppressing agents etc. to create a sense that something was being done and there is something on offer. This left me reaction less; but this is what generally happens when we treat diseases without any specific treatment. It wasn't anything new to me and not unexpected as well, but still it left me a little disappointed. Sadly some patients with Dengue also lost their lives in the Intensive Care Units, which made more news than all that I have written before. And I knew this will continue to scare everyone, the community, the doctors and the health care authorities. Coming year may not be any different, but I thought neither would I be.

HOPE FOR SURVIVAL

I had lived my childhood in a small Indian town, and in the eighties when communication and connectivity was limited, one actually lived a life isolated and insulated from the world outside. All my imagination was moulded by what I saw around, I had developed my own ideas and perceptions based on people and things around me. So was my idea about a doctor; it was limited to the very few I saw and met in my hometown. There was no qualification attached to them; a doctor was a doctor, except for the broad category of a surgeon or a medicine specialist. And I was fascinated by my family physician, that one person who seemed to have a solution to all our medical problems. The one person we approached for every illness, the one person for all age groups and both sexes; for me he epitomised a healer. He was a larger than life figure; being a doctor felt like a dream which I wanted to realise, except that it appeared unrealistic and distant. It reminds me of the story of the village schoolmaster, where everyone in the village marvelled at his knowledge; everyone was so fascinated how one brain could hold so much inside it. My feelings were very similar. Even today the reverence attached to a doctor is way above what any other professional could command, but not everyone is able to earn that. Today is the world of specialisation, of information and competition; being a doctor is so much more difficult and complicated. And your ordeal doesn't end with earning a qualification (which itself takes an epic time); you have to continue to go through the grind to be recognised and established. It only gets worse if you are a general physician, especially if you are a small person in a big city. I never knew these equations; but it did not take me long to realise that survival was tough. I was having my blues, I needed something to get out of it and thankfully it happened, bringing in a hope for survival.

I had completed my graduation which awards us the MBBS degree and then went on to do my post graduation in medicine which gives us an MD degree. I had already spent 11 years in the medical school to achieve that (with almost two years wasted in academic delays), had been married and almost hitting my thirties. I did not want to spend more time in education; in fact I could not afford to! If so many years of learning were not enough, then it never might be. And although I had realised that it was the age of speciality by the time I was to exit my medical school, I still could not wipe off the image of a general physician I always wanted to live. Except that this was in a new age, truly in a new century.

After my post graduation, it was time for me to start working. The only thing I knew was that I did not want to work in my medical school hospital; but I had no clue how to start off individually. I was worried I may have no work and a lot of time; worse still was that I may have to keep running around, with neither time nor money! Doctors can have a very tardy and trying start to their careers and to save myself from that, I started part time as a salaried professional in a tertiary care hospital. There you feel dwarfed by the presence of the big names around; to survive in that environment I had to acquire an expertise. Thus grew my interest in critical care, the Intensivist was slowly being recognised as a specialist, and the best part was I did not need any additional academic qualification to be one. I needed a little bit of training, a little more expertise and a little understanding of the workings of a critical care unit. It was like a little extension to my qualification. Luckily I could manage that in just a few months; that first fear of getting lost in oblivion was taken care of.

Although I still needed the skills I had acquired during my years in medical school, being in Intensive Care Unit was a different ball game. I had never imagined myself facing life and death situations daily; early in my career I had written more death certificates than prescriptions! Medical emergencies never come with appointments; so I had to be ready to work anytime, any day. Night and day, national holidays, festivals; diseases have continuity just like life! And we are a unit at work nonstop; just like our heart beat, meant to keep other hearts beating! It looked a little odd as a physician to be in the ICU, not only to others but sometimes to myself as well. But somehow I developed a liking for my work and my interest was infectious; four more of my MD colleagues joined me and we are now the big 5 together. May be it is this company which kept me going. But I was still a split man, I continued the rest of the time I had in practising general medicine, the one thing from which I could never take my interest away. Childhood fantasies seldom die; I had to live the role of a physician. But a year on and it was all disappointment. Neither was I known as a person nor was my qualification enough to make me known. In fact most people came to recognise me an Intensivist but it was going to do no good to my Out Patient general medicine practise.

I had heavily bet on my abilities to survive (actually thrive but now it had come down to mere survival in a year) but the health care system is such that one can get easily lost in it, not only because of the intricacies and uncertainties of medical science but also because of the kind of health care delivery system that we have. People need special solutions for simple problems; the focus is generally more on the doctor than on the disease and we have a population driven by brand image. I was like a non entity, wondering will there ever be a place for a family physician; that one man who has solutions for most health issues, the one man who you need to have as your health manager.

I was clouded with these thoughts when an elderly gentleman walked into my OPD. He appeared to be in his sixties and was accompanied by a young man who seemed to be his son. We had a leisurely start to the conversation and I came to know that they had learnt about me from a family friend.

"My brother in law told me about you and I thought of consulting you for my father," said the young man, my assumption that he was the son was right.

"My father has been having a lot of health care issues and we now seem to have a problem not only because of his diseases but also because of doctors. May be you could be of a lot of help." His statement was difficult for me to appreciate, but the fact that he started off with concerns of a different nature made me curious.

"I would like to know his medical history," I said, since I had been seeing him for the first time. I was handed over a voluminous file; it must have been as bulky as my textbook of medicine. And as I flipped through, I realised it also had the content of a lot of chapters! Seeing me lost in the vast sea of information, his son came to my rescue.

"My father has been having diabetes for the last twenty years. Of late he has developed complications. He has developed kidney dysfunction but does not require dialysis, he has Ischemic Heart disease which is being managed on medicines and he recently had an angiography but there was no need for intervention, he has disturbing paraesthesias which has been said to be due to diabetes and last few weeks he has been having persistent diarrhoea." His son really made my job easy. Now I could read his records with better focus. The fact that his son could give such a good summary (with technical terminology) was because his health care issues had been well taken care of. He was being seen by doctors of great repute, all of whom I knew personally. He was also under regular follow up; an endocrinologist was taking care of his diabetes, a cardiologist was looking after his cardiac problems, there was a neurologist who had been seeing him for his neuropathy, a nephrologist for the kidney dysfunction and a recent prescription of a gastroenterologist for his chronic diarrhoea. I was wondering what had brought him to me; it seemed difficult for me to find a place for my prescription in the file that he was carrying. I had to ask.

"What has brought you to my clinic?"

"You need to take care of my problems," this time it was the patient himself who spoke. He seemed weary and his response was like a reaction to some sense of disenchantment.

"They anyway have been well taken care of," I said. "Is there a specific reason you are here?"

"Yes. I need a one point solution for my problems. It has been difficult for me to get across to my doctors; the appointments are difficult to get, I do not know whom to approach for which problem of mine, sometimes there is an overlap of prescriptions which gets confusing and I am afraid the list of doctors may get longer."

"We had four appointments in the last 7 days and I am yet to get one for the gastroenterologist for a review," his son said. "The diarrhoea may have been prolonged because the treatment for it is yet to start. If you can manage his health, it will make things easy."

Although he had multiple problems of serious nature, his treatment was still about simple basic things. And one aspect of it was also making health care easily available to him. I have always believed that problems could be of complex nature but generally the solutions are of simple nature. Here perhaps the solution for him had become complex. I was also reminded of the fact that a patient is an individual and not just a collection of diseases; the treatment has to focus on the individual as well and not just on the diseases that he carries. The way he had been treated may have been ideal but it was not seemingly practical. He needed a health manager and suddenly I was rediscovering the role of a general physician, a family physician. I was reminded of the one man I had admired who had made our lives so easy. May be I had an opportunity to live that role.

"I am expected to treat the problems your father has," I said. "I can also liaison with other doctors if need be for his management." The latter was a diplomatic statement. I knew his father's condition would worsen with time and there was a possibility that I might be considered the reason for it rather than the nature of the diseases. That is human nature and it is best reflected in chronic progressive irreversible diseases. Patients keep juggling between doctors (even the specialists are not spared) in the hope of a magical solution; this hope is even essential to help them live through their diseases. Most people get the strength to bear the pain and the agony of chronic diseases (and even their treatment!) only because of faith and hope. You take that away and survival can seem deeply depressing. I always had to keep the hope alive; and also honestly try to give him the best of what was available for him.

"That would be perfect," the patient said. "You can go ahead and plan out my treatment."

It took some time before I could figure out the different possibilities of the diarrhoea that he was having; if that could be treated it would establish the much needed bond between us. The first meet ended on a promising note, bringing the much needed cheer in both our lives. For the first time, I realised the logistic requirement of a physician. My patient had rediscovered the way he had to be treated and I rediscovered myself. The meeting had brought in a hope for easy survival, for both of us.

A WORD TOO LATE

Sometimes a word of advice comes too late in the time span of a disease, sometimes it doesn't come with the force and emphasis that will drive home the point strongly. Sometimes it doesn't come from the right authority and sometimes perhaps it isn't cared by the patient, especially so with chronic irreversible medical disorders. This results in a situation where the patient goes from person to person and is always hunting for somebody who says something positive about his illness, someone who builds up hope of a possible cure and his focus changes from the actuals of his disease to actually what he wants to listen. Thus an opportunity for timely treatment is missed and many a times when it comes to health, irremediable damages are caused. And in the Indian health context the damage is not restricted just to health, it's also to wealth, profession, family and may be more than I can imagine. The following story deals with one such incident.

It was still early in my career and there were few patients, far in between (may be that has not changed dramatically even now!) and hardly few people knew me, even in the vicinity. Just midway into my OPD timings (but in terms of my patient the first one) a couple entered my room. I had never seen them before and they too knew nothing about me except for the fact that I was a medicine practitioner. The gentleman was well built, looked like in his late forties, had a muddy complexion and coarse dry hairs and he was the one who needed a consultation.

"Good evening Sir", the lady began, "He is my husband and we have come to discuss about his health issues. The women seemed more troubled than her husband, who appeared a little slow and emotionless.

"OK', how can I help you?"

Before I could speak further, she took out a bulky file. "These are his medical records. We have been told that he has liver disease and now his problem is worsening."

I wanted to hear more from her but she took a long pause, perhaps meaning that I should first see his medical records and then resume the conversation. I was worried at the sheer volume of the file but it did not take me long to understand what was wrong. There were prescriptions and laboratory reports all pertaining to his liver disorder and was more or less a monotonous repetition of some simple major facts. These carried his medical status of around two years and bore the signatures of many doctors all over the country. They had been to all the big names, not only in terms of doctors but also hospitals and places. I could make out easy and fast that the patient probably had developed cirrhosis of the liver, a chronic, progressive, irreversible cause of liver dysfunction which leads to end stage liver disease and death. What can possibly be done is delay the inevitable and ensure for a good quality of life by judicious and simple but meticulous management. A liver transplant is the only hope for long term survival.

Fortunately it had started in the country but unfortunately it is not easy to get one done. It's a challenging logistic in every way; few centres, paucity of cadaveric organs (one mostly has to get a live donor to get a transplant done which may sometimes be impossible due to non availability of a suitable donor) and a huge financial drain. I do not know of any free or even subsidised programme of liver transplant and the costs can be beyond the reach of most of the patients with cirrhosis.

The first thing after seeing the records which came to my mind was why the couple had come to me after having seen the big and the famous all over. One possible link which I could see was that they lived in a neighbourhood colony. Thus I could have expected a casual visit for a fever or headache but discussion on irreversible, lethal organ dysfunction and that too to their neighbourhood non entity (which I thought I was!) was something which I wasn't able to come to terms with. I might get to know possibly very soon. The other thing I realised was that the man was still 39yrs and I had overestimated his age by almost a decade, perhaps that's the effect the disease was having on him.

"I have thoroughly examined his medical records (must have been lightening fast the lady must have thought) and learnt that he possibly has cirrhosis of the liver. His virus screening is negative and there doesn't seem to be a specific cause for his liver dysfunction. Had he been taking alcohol?" This query got the atmosphere charged.

"No Doctor, he has never consumed alcohol, not even for recreational purpose. Forget alcohol, he doesn't even chew betel nut". (That's a common way of stamping the teetotallers' tag on a person in Eastern India where chewing of betel nut and tobacco is very common). He has been a committed family man, very hard working and has always consciously kept himself away from bad company. I just cannot accept the fact that he has developed cirrhosis of liver."

By now I came to know the reason for denial of the disease. True, alcohol is the leading cause of cirrhosis in our part of the country, may be more than 50%, and is followed by chronic viral hepatitis among the known causes. But what is largely ignored is that almost 25% of patients have cirrhosis of unknown cause, a majority of which are considered to be due to Non Alcoholic Fatty Liver Disease (NAFLD). This patient fitted into this category. Even as physicians we are reluctant to accept this diagnosis, I can understand how difficult it must have been for this couple. But I don't think they had come to me for this reason.

"What problem is he facing now?"

"He is having worsening swelling of his legs, also he has easy fatigability, forgetfulness and poor concentration," the lady said.

"These days it has been difficult for me to look after my business", I heard the male speak for the first time.

"Keeping calculations, taking judgements, staying alert has been difficult for me." His description was clearly suggestive that he had developed hepatic encephalopathy (brain dysfunction secondary to liver disease). This was still in its subtle form, where the patient may appear fully conscious but they may defects in their mental function. In later stages, it becomes more overt, with disturbances in their sleep pattern, levels of consciousness and behavioural abnormalities.

I asked him to come to the examination couch. The swelling in his legs was clearly visible; the legs were grossly swollen, making it difficult for him to wear even his footwear. On closer inspection, I could see the skin had become heavily pigmented, thick and coarse, suggesting that the oedema (swelling) was longstanding

"Your leg seems to be poor condition. I am not sure but there could be superadded infection of the skin. How long have you been having this swelling?"

"For almost two years now. It just does not seem to subside."

"What medicines have you been taking?" I wanted to see what medicines the patients was taking and was he aware of the steps to be taken to keep the swelling controlled. A look at his prescription made me realise that he had been on low dose diuretics (medicines which clear excess fluid from the body).

"Haven't your adjusted the dose according to your swelling? Do you check your body weight regularly?" The patient wore a dazed look. Perhaps nobody told him. A simple word of advice was missing from the not too simple treating team. In these conditions, we generally advice the patient to check their body weight regularly which is a sensitive and simple means of estimating excess body water and ask them to adjust their intake of salt and diuretics accordingly. At some point in the disease, even these measures fail, know as diuretics refractory edema, but they should be at applied in the first place at least! Due to suboptimal treatment, his leg skin had been irreversibly damaged due to long standing oedema.

I had to restart the conversation. I thought first I had to know what they knew and then may be make them understand all that they need to know.

"Do you have any idea what is the natural course of this disease and what are the interventions available for treatment?" I did not want to use the word transplant in the first place and that too so early, may not go down well with them.

"This time we have been told that he might require a liver transplant", his wife replied. Good that at least they had heard about it.

"Yes. That is also the advice that I would give you. There are few transplant centres now in the country and you can go to one of them. The process is fairly successful and since he has only liver as the dysfunctional organ and he is pretty young, he should do well."

"How much would it cost?"

"Around 20 to 25 lac rupees, I suppose."

The lady nearly broke down. "We have already spent that much in the last 2 years, travelling all around and getting the best possible treatment for him. His business has been affected because of our travel and in recent months because of his disease. We do not have a strong logistic support and I don't know how I can manage that much of money."

This was too much to spend to know that one has cirrhosis of liver and needs a liver transplant, I was telling myself. And look where the disease has brought them to, to their next door physician! Now I knew why exactly they chose to come to me, the disease had now also become a huge financial burden and the treatment process was just like a ritual that they had to perform. Now that they had nowhere to go, they finally accepted that they needed a transplant which they anyway cannot afford. Sad. But they would not have accepted it two years back if they had heard of it from me then. Treatments of such monumental nature are accepted only if the advice comes from the Tsars of modern medicine and that too with a force. May be that had not happened. Or maybe there was a constant denial of the disease which made them seek one doctor after the other. From the human angle, it actually might be difficult for a teetotaller in his late thirties to accept that he has developed cirrhosis of the liver which would ultimately lead to liver failure and death, that too at such an accelerated pace. Whatever it is, it was disheartening to see a young man losing his life partly to cirrhosis and partly to our health care system. He could never receive what he deserved, not once but on many occasions. Somewhere I felt even God was unfair, but then I thought that's how life is. What bigger misfortune can be than the loss of a life; working in an Intensive care I am a regular witness to it but one can feel the pain and agony every time it happens. With my involvement in the

Intensive Care, sometimes I feel I have written more death certificates than prescriptions! This time I was not signing on a death certificate, but neither had my prescription any life to offer to my patient.

NOT ANOTHER DAY

The Intensive Care Unit is always a high drama place; it has to be when it's a question of life and death. In the beginning it was difficult for me, since during periods of grief the emotions are very intense and so is our involvement, both with the patient and with the family. Loss of a precious life in such situations made me very uneasy; I almost became a member in the bereavement. It took me a lot of time to become immune to this viral feeling, may be experience also moulds our reactions to situations. On the flip side, it also moderated the highs that I felt when patients made remarkable recovery. It was almost as if I had accustomed myself to a roller coaster ride and although it still thrills, there are no after effects. But this element of intensity is something that I wanted to stay; otherwise my work will become a daily monotony and a drag on my energies. Perhaps it was time for me to relive the high drama again, this time even more intense than what I have ever experienced. I realised that however experienced I may be, when it's a dear one involved, I also react like every other person. This time it was not another day, because it was not another person. It was my junior colleague in the ICU, not as a doctor but as a patient, and he could not have been more sick. It was truly a dramatic event; we were lucky to see a better day.

It was around ten at night and it was time for me to retire. Just then I got a phone call from my ICU and I rushed out of my bed, sped to my car and off I went. My wife was curious because I had not spoken a word and the urgency in my activity was alarming. She gave me a call. "I am going to the ICU. Something is seriously wrong and I got to be there. Do not wait for me to come back. I think I will have to stay back." Whatever be the nature of urgency, it never happened before that I had to rush back to the ICU, some one or the other always managed. Ours is a team effort and every one contributes to the treatment process. But the event was unprecedented and my colleague from ICU could not help but call.

The traffic was thin and so were my spirits. I dashed into the unit and found my fellow Intensivist. "He has gone for a brain scan. I saw him when he had come and looks like he has suffered a stroke. The CT scan report should be here in a minute."

This was unexpected. He was still in his early thirties, two years younger to me, did not have any disease which I knew of and to have a stroke at this age is not something that we see usually. The scan plates had come and so had my colleague. I first had a look at him. He was drowsy and responded to verbal commands. There was no spontaneous eye opening but he opened when trying to respond. He was moving his right hand but there was no movement of the left side of his body. I immediately had a look at the scan. He had suffered a massive intra cerebral bleed. My heart sank. The sheer size of the bleed and the distortion it had caused to the brain parenchyma was huge. It could be appreciated by anyone who had a look. I was fearing the worst. The only silver lining was that he had survived the event and had reached the ICU. The bad news was that the damage had already occurred and there was no specific therapy for control or reversal. We were at the mercy of the Almighty and were praying for survival, the issues of long term disability did not even cross our mind.

I turned and saw his wife. She had understood, but perhaps she could not appreciate the extent of threat he was in. "He just drove back home, had his snacks and said he was not in the mood for dinner. He was relaxing on his bed when suddenly he had severe headache and started feeling unwell. After a minute, he developed nausea and vomited profusely. Before I could realise what had happened, he fell down in front of the mirror and became partially unresponsive. He tried to speak but he could not. I knew something disastrous has occurred and my worst fears are true."

I actually did not know how to respond. Was I supposed to console her, to support her, to explain her, she was herself a doctor and her husband not only happened to be a colleague but now he was also in our unit as a patient. With a stroke for which we did not have any answer! I immediately turned my attention to his progress sheet which records the vital parameters and the treatment plan. There was not much that I wanted to know, but that was the only way to keep silent and be involved. Words were difficult to come. Just at that moment our neurologist arrived and it did not take her long to realise how bad the situation was. By this time his father had also arrived. It was time to talk to the family. My neurologist gathered the nerves to speak.

"He has a haemorrhage in his brain involving the right basal ganglia and adjoining areas," she displayed it on the scan and it really looked terrible. "The volume of this would be more than 40ml which is considered large by our standards and there is also a shift in the brain structures due to the bleed." Brain is contained in a close tight space which our skull is and any increase in volume leads to distortion of the brain structures. There can also be a sudden death due to dysfunction of our vital centres. "The bleed has also extended into the ventricular system which further increases the risk." It was a total technical description. And appeared ominous! Some hope had to be given to keep the conversation going.

"His level of consciousness is what which gives me great hope". This assessment did not require any technical sense, although we score it based on the Glasgow Coma Scale. "What would be most important is how it stays over the next few hours because the bleeding can keep expanding for the first 24hrs." This meant things could worsen from here. May be he stops responding to verbal stimuli. May be he does not maintain his airway and has to be intubated and started on artificial respiration. May be...... there were so many things which could go wrong. "There is very little that can be done in terms of surgical management for hematoma removal, there is no evidence of any benefit and I have my doubts if any neurosurgeon would actually like to do it. A decompression surgery may be considered whereby a part of the skull bone is removed to give more space to the swollen brain."

My neurologist had put it across very impersonally. The facts had to be told and she did it well. But it almost appeared that we could do nothing and had to wait for the tide to turn. It was a helpless situation. Two more of my Intensivist colleague arrived and it appeared the system was activated. It would be stupid to sit back and wait, let's work out all possibilities tonight, tomorrow may be too far away, I told myself.

"We need to get the neurosurgeon to see him right now", my colleague said, almost giving a purpose to our presence. 'If he deteriorates, an emergency decompression surgery may be in and it's better if he has a look right now." It took another 15 min for the neurosurgeon to reach. The easy traffic and free time made every one reach instantly.

But no clear cut conclusions could be reached. That if we needed to wait or may go ahead with a planned decompression surgery early in the morning. I had never bugged my brain on this issue before and never had an Intra Cerebral haemorrhage made me think so much. We decided to take another opinion. The discussions carried on till late midnight. We had to be sure that we did not do any further harm, in the expectation of some magical turnaround. But may be out of sheer helplessness and the compulsion to do something, we decided to proceed with a surgery early morning.

The night was uneasy. Both about what had happened and what we were going to do. And in between there were flashes of the bygone days. He was a livewire, it was impossible to believe that he was bedbound with a stroke and may remain so for long; it could be even worse. He spent a lot of time with us and if it was another day, he would be around, fiddling with his I-phone or his pad and still ready to rush to any emergency call. With him around, handling any emergency was much easy; we were now caught in one in which he could not help! Everybody around was having similar feelings but hardly anyone talked. Everything around was still; somehow there were no other emergency that night and this made it even more perceptible. It was one of the most helpless situations I was in as a doctor and I was desperately trying to come out of it. But nothing seemed to help; neither medically, nor emotionally. But I knew time was ticking. I wasn't sure if it was taking us away from a disaster or towards it. There was nothing wrong till now but a lot of it had already happened. We were not anticipating any dramatic turnaround and I was wishing my services would not be needed as an Intensivist. This can be a sick feeling; where the wait and watch was only for things to worsen, where every monitoring method was applied to detect any worsening, where every time I had to go to him to see that he was all the same. It was a drag on my mental energies and I was fatiguing, my body was used to this but not my mind. It was almost the sunrise the next morning and I was getting dizzy. We were all in our room in the ICU complex when another fellow neurologist came. He came to know late and rushed early in the morning to have a look.

"I am still uncertain about the benefit or the actual need of a surgery. But since you all and the family have decided, you may go ahead with it."

Although I might have been in a state of trance, I overheard everything very clearly. I immediately woke up and started," There is no final word as yet. And the family is in the least capable state of taking a decision. It is we who are taking the decision and they have actually left it to us to do what is best for him."

My neurologist was startled at my response. He had expected that I had slept off and was about to leave. Now that was not to be. He had to be a part of the decision making process. Both of us decided to have an extensive discussion with whoever we could catch that early in the morning. Another hour and we decided against the surgery. Now our focus was on regular assessment, avoidance of any iatrogenic insult and a wait and watch policy. It was almost 24hrs and nothing had gone wrong till now. This was a sense of great relief. But the fear stayed. I was supposed to do the night duty but I refused to stay alone.

"I do not think I have the nerves to stand it all alone. Although he is doing well, I may start feeling sick if I am alone." For the first two nights, it was two of us every time. The atmosphere in the ICU was sedate and sombre but everyone was trying to put up a brave face. That was essential to send the right signals to the family. It also made the discussion a lot easier to come.

"He had not been taking care of his hypertension. He did not even listen", said his wife. I had never known that he had hypertension. In fact he never talked about any medical issue of his, although we spent a lot of time together discussing the medical issues of our patients. I also came to know that he was a pampered child who had come late after the marriage of his parents. And that he was the only child. It suddenly reminded me of his parents. They all needed him so badly.

"I want my son alive. It does not matter what disability he has later," I remember his father saying. Even we were bothered about the same issue. It was good that more than two days had gone by and he remained stable. Statistically he had an almost 50-50 chance that he recovers; his ability to walk independently was still lesser. That was threatening but we still had a good chance. And I was looking at the positive side. He had a bleed involving the right side of his brain which is the non dominant hemisphere in more than 90% of patients and so was in his case. This meant that his ability to understand language and to talk would not be affected. Perhaps that is why he was still responding to commands and making an effort to speak. In fact we could hear a few words from the second day. This was of paramount importance to both his personality and profession; he would not lose his medical intelligence. A stroke on the right side also meant that his left half of the body would be paralysed. His right hand would be fine, he should be able to do all the skilled job of a single hand, the most important of which was his ability to write. I was stirred; if we could save him, he could be back to most of himself.

After the first four days, there were signs of recovery. He became more alert, although he still used to feel sleepy.

He had fewer headaches and lesser need for analgesics. He also started to eat. He was so fond of eating, I remembered. Even I am a foodie and it was like a food festival when he used to be around. Night duties when he was on his emergency days was so much fun, you could see him full of energy at any hour in the night. The fact is I always used to tease him at the end of duty hours. You always see people getting ready for their duties but with him he got into his preparations when the duty ended. He always had somewhere else to go; he had so much zest for life.

Towards the end of the week, we knew the worst was over but we still were not out of the woods. The best part was that intellect was fully preserved and now he could speak much more clearly.

He could not only speak but also exercise his medical intelligence. He had complained of chest pain when I was on duty and I immediately ordered for an ECG.

"This is not cardiac pain," he yelled out at me. "This feels like muscular pain. Please give me an analgesic." The orders were immediately followed. I was thrilled at his participation. And he was absolutely right. The ECG was normal and his pain improved with analgesics.

He also has recovered his limbic senses. I had once got ice cream for him, because we often had it together otherwise. "Would you like to eat ice cream?" I asked as if I had not got any. "Yes, get me a Baskin Robbins fresh fruit flavour."

"Actually I have already got the another variety." But he was bent on his choice. The ice cream parlour was half an hour drive from the hospital and I had to get it. And

I got it for him after I was done with my duty. It was a moment of joy.

It took almost three long weeks at the hospital before we could send him home. It was a great sense of relief but it was not as much joy because he had to go back on a wheelchair. We actually wanted him back on his legs, back in the hospital, this time not as a patient but as a doctor. That took three months, but he was back. Back as the Emergency Officer! He still had residual weakness on the left side, but mentally he was now even stronger. He is back fully as the doctor; maybe he is yet to come back to his full life. When I look back at the whole episode, I do not know if I should feel disappointed or blessed; what happened was preventable but it could have been even worse. Whenever thoughts of that day flash in my mind, I feel relieved that it is over. And although I always look for thrills and challenges, I pray to God that I do not have to live one another day like that.

WITNESS TO THE NATURAL HISTORY

Most of the chronic diseases have a natural progression of events, which can be defined into various stages and time frames. Generally they have a latent asymptomatic phase which can last for years, and then organ dysfunction develops, which initially may be asymptomatic or produce some symptoms, and then progressive organ dysfunction leading to organ failure. Although predictable, this is not an absolute truth; some patients despite the presence of risk factors and diseases do not have any obvious impact throughout their life and die due to unrelated causes. Thus one may smoke a whole life and never develop lung cancer, one may have uncontrolled diabetes but still remain naturally protected, people live even into their eighties and nineties with high blood pressure and continue to enjoy good health; but the problem is one does not know who will be spared and who pays the price. So individually we may not be able to predict the health benefits of treatment of these conditions, but at a community level it is definitely beneficial. This is the essence of preventive health care; identification of risk factors by screening during the latent asymptomatic phase and initiation of treatment to prevent progression and organ injury.

We get to see this progressive downhill course in so many patients but one particular patient was of special significance; probably because of the way the sequence of events spanned out. It was also a classical example of how a preventable disease went on to end organ failure, not because of lack of health care logistics but because of poor understanding and denial.

I happened to see a 62yr old gentleman in my OPD who had come for a health check up. We provide a set of tests as a package and people choose this as per their liking (which is surprising as well as undesirable!) and generally meet me with their test reports. He was lean built, agile appearing and during the interview I came to know that he had been into a busy business schedule. He took out the time for a health check up only on the insistence of his son. His son was also there by his side. There was not much to enquire because he did not have any symptoms. I just had to flip through the reports. They were all normal, which my patient also knew (abnormal results are always highlighted in our laboratory reports); he came to me probably because it was a part of the health check to meet the doctor once. It appeared he had very little interest in the medical interview, which must have been reduced further seeing a young doctor on the other end!

It was time to do the physical examination (things were occurring in the reverse order; the investigations first, the physical examination next and the medical history was left to the last!). His body weight was on the lesser side but not of any significance; the entire examination was unremarkable except for blood pressure which was very high.

"Your BP is very high, its 180/110mmHg when ideally it should be less than 120/80mmHg," I said with an emphasis on the figures and creases drawn on my forehead, trying to convey my sense of concern.

"Oh! At times it becomes high, you know all the stress at work," he said nonchalantly.

"That means it has been documented to be high earlier as well?" I enquired. He had denied any pre existing medical problem. (Perhaps he didn't consider it as a problem!).

"Yes, but that is during stress. It has never caused me any problem. I know this will settle down on its own."

Blood pressure can increase during stress due to sympathetic over activity, but they are not mutually exclusive. You can blame this on the nomenclature; hypertension and blood pressure has these two words tension and pressure within them and to the lay man it always implies as having some strong relationship to physical and mental stress! Add to that the asymptomatic nature of the disease and the need to take long term medicine for control; denial is the most obvious human reaction, even in this age of information. My patient was no different; for him the pressure will go away with the stress. He couldn't realise how much more stress it was going to bring! But as the physician it was my duty to make him understand the importance of the condition and its management.

"We generally do not make conclusions based on a single blood pressure reading but when it is this high, the chances that it will normalise on its own is very little. If uncontrolled this can potentially lead to organ failure, most importantly the heart, the kidney and the brain". I handed him over an information brochure on hypertension which had been specially prepared for patient education. "You can read the literature yourself; this has been prepared from standard international books." Sometimes the communication with the patient may fail to drive home the point; we may not be able to give enough time during the interview and the patient may still have queries in his mind which may be of conflicting nature. A handout in such conditions is something I have found to be of help; it also takes care of any omissions. And when you say international, the patient is more likely to value it, rather than the words flowing from a young inexperienced doctor.

"But there has not been any damage till now, all my reports are fine," the gentleman said, not convinced that anything wrong could happen to him.

"By the time the reports show abnormality it's too late, we have to work in anticipation and ensure that the reports always remain normal. The surest way would be to keep the blood pressure in the normal range."

"OK doctor, I understand your point. I will see you again very soon." The intent was missing from his voice. What he probably wanted to say was that he was least bothered about whatever I had to say and it was time he left. It was a dignified way to end the conversation.

I hardly had any time to write a prescription but during the interview itself I had jotted down the main issues and I handed that down to the patient. The most important thing for me in the prescription was that it carried my name and contact, in case there was a change of heart (or may be disease of heart!). A few warm words of exchange and the meeting ended.

As expected, he did not turn up. I was hoping he had found someone else to take care of him. Looks like that did not happen. But he surprised me by coming back after a full two years and even more surprising was that he was still carrying my prescription!

It was a repeat of events. He again got few tests as part of his health check and came to me with the reports. At the first instance I could not recall his face but a look at my prescription and everything came back to my mind.

"Remember we had met two years back?" he said with a leisurely smile. "I just thought I should get a review. I don't have any problem but one of my reports is highlighted. Can you tell me what is to be done?"

"Have you been consulting someone and taking medicine for your hypertension?"

"I did not feel the need."

I glanced at his reports and now I knew he will in the need for a lot of things! His creatinine levels had increased which signified kidney dysfunction. I knew the statistics had caught up with him; unluckily he was the one who had to pay the price for the neglect.

"I am sorry but you have damaged your kidneys. The process cannot be reversed and the deterioration will be progressive, except that if you take good care even now, the progression can be slowed", my tone must have been pessimistic but that was because I knew he was in for bad days ahead.

"My creatinine level is only marginally raised. It is 1.9meq/L whereas the normal is up to 1.4meq/L. Is it that concerning?"

"Yes. More than 50% of kidney function has to be lost before the creatinine level starts rising. Once kidney function deteriorates, it becomes self perpetuating. The BP also becomes more difficult to control." But even more important was the fact that he was a difficult to control patient! "I am afraid the window of opportunity to treat you has been lost, but still the one thing that you have to do is aggressively control your blood pressure. Good care even at this stage will extend the good days for you."

"And what about the kidneys?" he was inquisitive, as if something had to done to get them working normal (or rather getting the reports normal. He had no problems; the problem was only the report!)

"There is nothing we can do to restore its full functioning; we need to focus now on damage control."

For the first time he appeared worried, but sceptic as well. He perhaps needed to find someone who could offer him more. But he did not hurry. I could give a full prescription; further diagnostic evaluation, medicines and a medical summary.

The parting ceremonials were missing. He took my prescription and slowly walked away. It took him a long time to leave my room, maybe he wanted to ask something or may be was disturbed; I could not make out.

The first stage in his disease had passed; I knew he was young enough to face the whole spectrum of the disease. 64yr would be considered a young age for a patient with a chronic disease. You may be expected to live may be a decade or more with the disease, the last few years of which could be really troublesome! This extension of life with disease can give you the worst time of your life. You struggle to move, you struggle to breathe, you spend time between hospital and home, your body starts decaying and you still keep surviving! Modern medicine makes you live through all of them, sometimes I wonder if it is a blessing or a burden. I was sure that my patient will have to go through it; what I was not sure was whether I would be a witness to it. But God had his plans laid it; and I and my patient were to spend more time together.

Time flew. It was 10pm; I was in the ICU, destined to spend the night in the unit. I was very fresh on my legs when a code blue alarm was raised. A patient was being rushed to the ICU. The stretcher came speeding inside, there was a patient almost gasping, it very well could have been his last breaths! He was immediately intubated and connected to the mechanical ventilator with high flow oxygen. He just made it, or maybe we just made it! It could have been a matter of only a few minutes! His pulse was full and galloping at 130 beats per minute, his BP was 180/90mmHg and he was sweating profusely. This was the most vivid picture of an adrenal rush; he probably had survived only on that. The chest was full of crepitations, it was very likely acute pulmonary oedema and the chest X ray showed that the lungs were flooded with water.

The crisis was controlled within a few minutes of initiation of mechanical ventilator. The turnaround was as dramatic as the deterioration, a few minutes delay and he may not have survived! The pulse slowed down and the Oxygen started rising; now I was feeling the adrenaline rush of a life saved! I could now have the time to look at the case records. He had a file with him, which thankfully was not bulky. I immediately went through the records. And among the few prescriptions, there were two of mine! I had an instant recall and I immediately rushed to see the patient. All this had happened too fast. He was just 66 and had his first episode of heart failure. The investigations showed that his kidney function had worsened, his heart had become thick and tired from the effects of hypertension and kidney disease and he appeared to have lost his muscle bulk further. What a chain of events I had to see; from only a risk factor to organ dysfunction to near death, I was a part of each of them! The journey from the OPD to the ICU had happened electric fast, and to me it appeared even faster because I never saw him in between. My mind was flooded with a lot of thoughts; it felt very upsetting that the patient had reached this stage. I was happy that I could save him but I was not sure if he would enjoy good health (and good days!) again.

By the morning the storm had fully settled. The patient had been discontinued of the sedatives and had become fully conscious. May be conscious deep down in his heart that he had been in a major trouble which may happen again! The fact that you breaths are being taken care of by a machine can be very disturbing; we live because we breathe and probably he had not been able to do that. These must have been moments of intense anxiety; the whole night he was fast asleep because of the effects of drugs and he could not have sensed it then. He could sense it now. It wasn't going to be long although, he was fit to be out of the ventilator and I went to extubate him.

"Look where have you landed yourself," I could not stop myself from saying that. I might have appeared sarcastic but I was deeply upset. This was the day I had warned him about and it happened, and it had to be me on the other side! He may have been surprised what I was doing in the ICU next to him, but very soon he realised that I was his attending doctor. Again! This time the treatment went off very well but only with good short term outcomes. I was sure I was going to see him much more.

Within two days he was out of the ICU and discharged from the hospital on the fourth day. In the last four years I had seen him twice but in the last four days I had to see him innumerable times. Now he would need more meticulous care, the disease intricacies would get worse, he may have to see a doctor, or may be many of them more frequently and he may still keep worsening. And he might lose all the fortunes he made for himself during those busy hectic days. Hospital bills can be a big dent on the purse, especially if the days have to be spent in an ICU. It was a battle lost on all fronts.

He kept on regular follow up thereafter but it was too late. I tried to bring comfort and quality to his life but the disease's behaviour was difficult to tame. A year later he became dialysis dependent and the ordeal ended a year later when he could not survive a massive cardiac event. That also happened in my hospital. But these were agonising three years which was very much preventable. He knew he had erred and even acknowledged it. "I did not listen to you," he said with a sense of remorse. And within myself I had a sense of disappointment; I wanted to treat him but not inside the ICU, I wanted him to be my patient but not this way.

THE MIDAS TOUCH

How about if you can relieve a patient of his suffering in a fraction of a second? Might seem unrealistic, but this is what this story is all about. The feeling is nothing less than magical and it is moments like this which reinforces our faith (as well as of the patient!) in modern medicine. Often we are found saying to our patients that we don't not have any magical cure, that the cure of disease takes its own time and one must learn to be patient; that sometimes we have nothing to offer except few empathetic words and some palliative medicines. This fact not uncommonly makes me feel how helpless we are. And especially in the ICU where we see people losing lives every now and then despite our best efforts, not to the natural process of aging but to the artificial state called disease which cuts shorts a life unexpectedly and untimely. It is in such circumstances that some magical moment keeps us propelled and charged. Being sheer witness to such an event is a transcending experience. And how about if it happens even without the use of a medicine, just a gentle touch or may be not even that? Simply unbelievable! And add to that a string of co incidences. You really have to be a part of that moment to feel the thrill and the joy. I would like to re create that scene and sense and hope I can make you feel a fraction of what I felt. If done, I would consider myself successful.

The scene is set in my ICU. It was 9:30pm and I had just landed for my night duty. Working at night has always been a different and special feeling for me and although it takes its toll, it's still something I might chose to do again and again. This nonstop 24hrs feel where every moment a team is working to save a life when everyone else might be in the cosy confines of their home calling it quits to the day is enough to make me thrilled. But what I was in for would only take the intensities further. Just when I opened the gates and entered the unit, I saw a crowd at the bed next to the door. I knew something was on and as a team member I had to join in instantly. When I neared the bed I came eye to eye with a friend of mine, who was standing beside the bed.

"Hi, how come you are here?"

I had known him only a few months now, more so because he was a patient of my wife and we had come across each other on few occasions because his office was next to my hospital. He was my contemporary, who had taken up the editorship of a newly launched English daily and had a typical make of a newspaper man; among commoners, politicians, socialites and always on for a cup of coffee. My reasons for getting friendly was that there were few people who value you as a Doctor if you look young (or relatively young, most of us are hitting our thirties by the time we start and isn't unusual for a father and son to be appearing for exams together, one for his masters and the other for his starters!) and he seemed to be one of them.

"My colleague at my newspaper office has been having palpitations for an hour now and when we came to the hospital, we were told his heart rhythm was abnormal and he would need admission in the ICU to relieve him of his problem. I am feeling mighty happy that you are around. I was here just to complete the formalities of the admission procedure"

"I had a look at the patient on the bed and he looked young, may be somewhere in his early thirties, not the age where people would have cardiac disease but a look at the cardiac monitor and the brief history given a minute ago made me know that he had Supraventricular Tachycardia. And I realised why so many people were around, he was being prepared for cardioversion."

Supraventricular Tachycardia is a condition where the heart starts beating at almost two to three times normal due to the activation of an ectopic (or simply abnormal) focus and becomes self perpetual. Thus our normal heart beat which is 60 to 90 per minute reaches around 180 to 200 beats per minute. The condition can start from nowhere instantaneously (thus it is also prefixed as paroxysmal) and produces a sense of pounding inside the chest. This can be worse than what you might feel after a 100m dash and there although after rest the heart rate calms down, here it continues to beat at a constant galloping rhythm. This disease can occur at any age group, is generally nonlethal unlike some other forms of cardiac rhythm disorders. It may last for minutes to hours or rarely even few days and terminates in some fortunate ones spontaneously or by the use of medicines or electrical shock, the so called chemical and electrical cardioversion respectively. My colleague had finished taking the ECG rhythm, confirmed the diagnosis and was preparing for chemical cardioversion.

The patient seemed anxious, now more so because of the people around and the beep of the monitor, which made him more aware of his every heart beat and he knew something was grossly abnormal. And when it comes to matters of the heart and you are in an ICU surrounded by a team in a distinguishing uniform, the fear and dread is understandable. I had never seen him before but he had overheard my conversation with my friend and was trying his best to pass a smile but it seemed a difficult facial expression to make. I was generous with mine because I knew the condition was treatable and it would only reinforce the confidence my friend has in me. My presence made my friend to stay back for a little longer and it was only a stroke of co incidence that we met each other because we generally do not allow attendants of patients in the ICU except for briefings. I reassured him that we will attempt cardioversion and that his colleague should be alright.

"You mean to say that he would not have to get admitted to the ICU and can actually go back home?" he seemed a little puzzled.

"Just wait", I was trying to be modest considering the uncertainties of diseases and medicines.

I now turned my attention to my fellow Intensivist.

"I have asked for Adenosine", he said. This is the drug used first up and can convert the rhythm to normal in an instant; the heart rate changes suddenly to normal, the conversion is abrupt without any lag phase or slowing down. The pharmacology of the drug is no less dramatic. It has a life of only around 9 sec in the body, has to delivered into a vein which is close to the heart so that it reaches there in less than that time and the mechanism of its action is still not known clearly!

"Did we try vagal manoeuvres?" I asked instantly.

"No", was his prompt reply. "Do you want to have a go before we inject Adenosine?"

"Not a bad idea," he continued. We can always wait for another minute.

Vagal manoeuvre is a method by which the tone of the vagal nerve is increased transiently. This nerve is responsible for slowing the heart rate (opposite to what adrenaline does) and can sometimes convert a Paroxysmal Supraventricular Tachycardia (PSVT) to a normal rhythm. I thought I would first try carotid massage. I explained to my patient that I was going to apply firm pressure below the angle of his jaw and this may cause mild discomfort but could have therapeutic value. I applied firm gentle pressure on his Carotid bulb for some time but it was of no effect. I thought of a retry but then I changed my mind and decided to try the Valsalva Manoeuvre, another way of increasing the vagal tone. Here, the patient closes his nostrils with a pinch, tries to blow out from his mouth but with his mouth closed (something which is also advised when we have water entering our ear during a bath, the attempt also opens the Eustachian tube).

I demonstrated it to him and asked him to follow. It failed the first time. Again and again but it didn't work. He tried it the fourth time and it happened! He had been able to do it all by himself. It did not require even the Midas touch! And it was all so obvious. The frantic beep of the monitor had suddenly died down and it was so easily appreciable. The ECG rhythm on the monitor suddenly changed and one did not know to read the ECG to recognise the changes. The complexes which were huddled together appeared distinctly separate from one another now. The heart rate which was flashing red at 190 per minute on the monitor changed to deep blue at 88 per minute and was no longer flashing. I quickly scanned the look on the face of my staff and they were all still, with awe and disbelief! Although they had seen a lot of reversals with adenosine but all that happening without anything at all was something they had not witnessed before. With Valsalva this was even my first experience, I had reverted one with Carotid massage earlier. But that was not as dramatic because it happened in the OPD, without any ongoing monitoring and other than the patient and myself, no one knew what had happened. But here it was like a theatrical act, with sounds and lights and an audience, with a well laid out script and a dramatic turn of events. The expression on everyone's face was unmissable, including my friend and my patient. He almost felt like a thud where his heart seemed to have paused, he no longer felt his heart beat (which is a normal phenomenon!) and suddenly he was feeling normal all over again. The chaos which his heart was in seemed to be over! And I was ecstatic. "Yes", was the loud outburst I made which brought that sense of stupor to turn into an outburst of joy and achievement!

But my friend was still in a state of daze. He, who had been busy getting the admission procedures couldn't believe that everything was alright. "What to do now?" he still wasn't sure if things would continue to be normal.

"Get me and my staff some coffee and get ready to take your patient home. He should be alright and in case the condition recurs, he can try the Valsalva or come to hospital". What we were all witness to was a miracle of modern science and I know there are so many more in store for us, I just hope I can be part of those events, if not the orchestrator, at least an audience.

Printed in the United States
By Bookmasters